Framework MATHS

HOMEWORK BOOK

David Capewell Formerly Westfield School, Sheffield
Marguerite Comyns Queen Mary's High School, Walsall
Gillian Flinton All Saints Catholic High School, Sheffield
Paul Flinton Chaucer School, Sheffield
Geoff Fowler Maths Strategy Manager, Birmingham
Derek Huby Mathematics Consultant, West Sussex
Peter Johnson Wellfield High School, Leyland, Lancashire
Penny Jones Waverley School, Birmingham
Jayne Kranat Langley Park School for Girls, Bromley
Ian Molyneux St. Bedes RC High School, Ormskirk
Peter Mullarkey Netherhall School, Maryport, Cumbria
Nina Patel Ifield Community College, West Sussex

OXFORD
UNIVERSITY PRESS

Great Clarendon Street, Oxford OX2 6DP

Oxford University Press is a department of the University of Oxford.
It furthers the University's objective of excellence in research,
scholarship, and education by publishing worldwide in

Oxford New York

Auckland Cape Town Dar es Salaam Hong Kong Karachi
Kuala Lumpur Madrid Melbourne Mexico City Nairobi
New Delhi Shanghai Taipei Toronto

With offices in

Argentina Austria Brazil Chile Czech Republic France Greece
Guatemala Hungary Italy Japan Poland Portugal Singapore
South Korea Switzerland Thailand Turkey Ukraine Vietnam

First published 2003

British Library Cataloguing in Publication Data

Data available

ISBN 9780199148912

10 9 8 7 6 5 4

The photograph on the cover is reproduced courtesy of
Graeme Peacock

The publishers would like to thank QCA for their kind permission to use
Key Stage 3 SAT questions.

Typeset by Tech-Set Ltd, Gateshead, Tyne and Wear
and Bridge Creative Services, Oxon

Printed in Great Britain by Bell and Bain, Glasgow

About this book

Framework Maths Year 8E has been written specifically for Year 8 higher ability students. The content is based on the Year 9 teaching objectives from the Framework for Teaching Mathematics.

The authors are experienced teachers and maths consultants, who have been incorporating the Framework approaches into their teaching for many years and so are well qualified to help you successfully meet the Framework objectives.

The books are made up of units based on the medium-term plans that complement the Framework document, thus maintaining the required pitch, pace and progression.

This Homework Book is written to consolidate and extend the Core objectives in Year 8, and is designed to support the use of the Framework Maths 8E Student's Book.

The material is ideal for homework, further work in class and extra practice. It comprises:

- A homework for every lesson, with a focus on problem-solving activities.
- Worked examples as appropriate, so the book is self-contained.
- Past paper SAT questions at the end of each unit, at Level 6 and Level 7 so that you can check students' progress against National Standards.

Problem solving is integrated throughout the material as suggested in the Framework.

Contents

NA1.1HW Adding and subtracting negatives

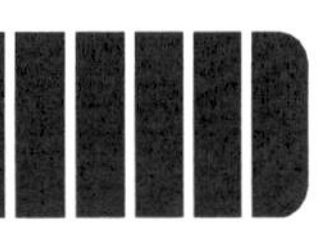

1 The table below shows the bank balances of five people.

Wendy	+£36.60
Patrick	$^{-}$£50.21
Carl	$^{-}$£0.37
Kevin	+ £5.06
Emily	$^{-}$£8.84

a How much do the five people owe altogether?

b How much more does Patrick owe than Emily?

c Carl has an overdraft facility of £100. This means that the bank allows him to borrow up to £100.
How much more could he borrow?

d Jane's bank balance is exactly half way between Emily? and Kevin's balances.
What is her bank balance?

e Write the balances in order, starting with the smallest. (Include Jane's balance.)

2 Copy these statements and write in the correct signs:

>, < or =

a $^{-}0.72$ ☐ $^{-}0.776$

b 3.1624 ☐ 3.2

c $^{-}5.6$ ☐ $^{-}5.60$

d 2.34 ☐ $^{-}2.35$

e 0.016 ☐ $^{-}0.0716$

NA1.2HW Multiplying and dividing with negative numbers

Remember:

× or ÷	+	−
+	+	−
−	−	+

1 Copy and complete this multiplication grid.

×	⁻5	⁻0.6	7
⁻2.5			⁻17.5
6			
9.9			

⁻2.5 × 7 (→ ⁻17.5)

2 Copy and complete these calculations by writing the correct number in the box.

a 2.1 × ☐ = 7.35

b ☐ × 0.003 = 0.0192

c ⁻0.24 × ☐ = ⁻108

d ☐ × 7.1 = ⁻5.68

3 Twelve friends agreed to share the cost of buying a minibus between them. They received a reduction of $\frac{1}{5}$ off the original price of £11 499.

How much:

a did each person save?

b did each person pay?

4 Here are two expressions:

a $x^2 + x - 3$ **b** $y^3 - 2y + 6$

Substitute in these numbers to evaluate the expressions:

i 3 **ii** ⁻3 **iii** 0.5 **iv** 0.35

NA1.3HW Multiples, factors and primes

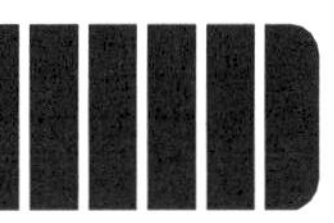

Remember:
- ◆ HCF is short for highest common factor.
- ◆ LCM is short for lowest common multiple.
- ◆ You can use prime factors to find the HCF and LCM of a pair of numbers.

1 Any even number is divisible by 2.
For example:

2584 is divisible by 2 because 2584 is an even number.
Describe the rules of divisibility for each of these numbers:

a 3 **b** 4 **c** 5

d 6 **e** 8 **f** 9

Give an example in each case to illustrate your rule.

2 Find the LCM of each of these pairs of numbers:

a 25 and 63

b 32 and 15

c 12 and 30

3 Find the HCF of each of these pairs of numbers:

a 65 and 23

b 52 and 35

c 56 and 18

4 Write each of the square numbers from 1 to 100 as a product of its prime factors.

Square number	Prime factors
1	No prime factors
4	2×2
9	
:	
100	

Write down anything you notice.

NA1.4HW Powers and roots

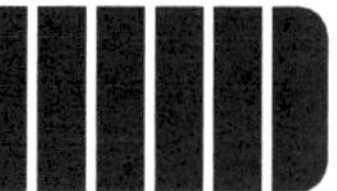

Calculators made easy!

Produce an easy-to-understand guide in plain English of how to use the powers and roots keys on your calculator.

Your guide should:

- Explain how to use these keys:

- Explain how to calculate the power of any number, including negative numbers and decimals.
  Remember to explain how to find cubes and higher powers as well as squares.

- Explain how to calculate the square and cube root of any number, including negative numbers and decimals.
  Remember, if you have a standard calculator, cube roots must be found by trial and error.

- Say what common mistakes can be made.

NA1.5HW Generating sequences

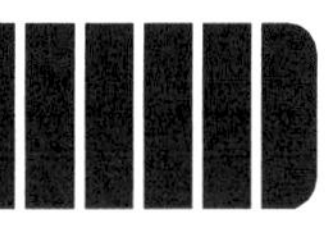

Remember:

- A linear sequence is a set of numbers that:
  go up by the same amount, 1, 5, 9, ...
  or go down by the same amount, 18, 16, 14 ...

1 Copy and complete these linear sequences.

a 5, 9, 13, □, □, ... **b** □, 7, 10, 13, □, ... **d** □, □, 87, 80, □, ...

2 Match each sequence in the clouds with its general term:

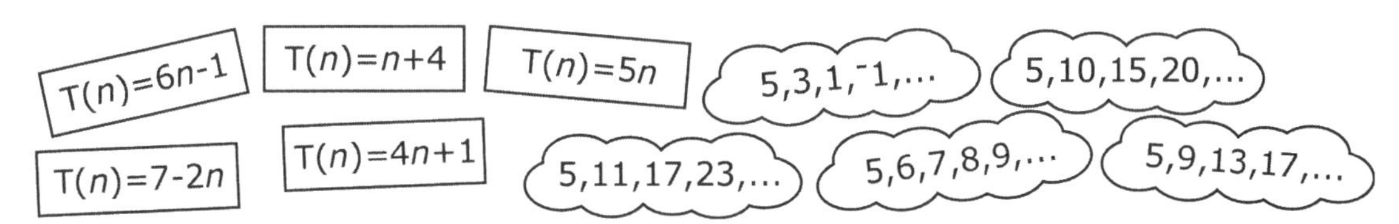

3 Find a formula for T(n), the general term, for each sequence:

a 4, 10, 16, 22, ... **b** 7, 10, 13, 16, ... **c** 2, 8, 14, 20, 26, ...

d 10, 8, 6, 4, 2, ... **e** 4, $4\frac{1}{2}$, 5, $5\frac{1}{2}$, 6, ... **f** 3, 2.8, 2.6, 2.4, 2.2, ...

4 A snail begins his journey 5 cm up a wall. He then climbs 2 cm every hour.

a Write a sequence to show his distance after each hour.

b Write a formula connecting his distance up the wall and the time, in hours.

c How far up the wall will the snail be after one day (12 hours)?

d The wall is 2 m high. When will the snail reach the top?

5 The 99th and 100th terms of a linear sequence are 351 and 353. Find, using algebra, the first five terms of the sequence.

6 Town houses are built with different numbers of storeys.

a Copy and complete the table of information:

Number of storeys (s)	1	2	3	4	5
Number of windows (w)	3				

b Find a formula connecting s and w.

c Explain **why** your formula works.

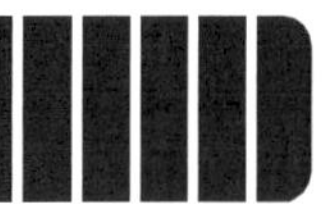

NA1.6HW Quadratic sequences

Remember:

- The general term of a quadratic sequence contains an n^2 term.
- To find the general term of a quadratic sequence look at the second difference.

1 Find the next two terms and T(n), the general term, of these quadratic sequences.

a 3, 6, 11, 18, 27, … **b** 5, 20, 45, 80, 125, … **c** 0, 3, 8, 15, 24, …

2 Generate the first five terms of the sequences defined by these formulae:

$T(n) = n^2 + 3$ $T(n) = 3n^2$ $T(n) = n^2 + 2n$ $T(n) = (n + 1)(n + 2)$ $T(n) = 12 - n^2$

3 **a** Copy and complete the table for each pattern.

i

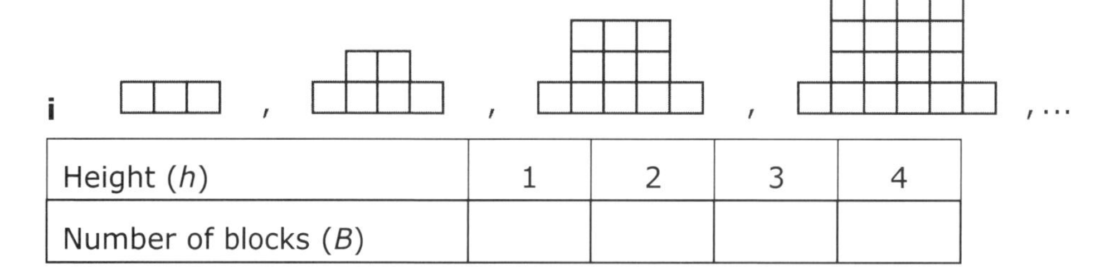

Height (h)	1	2	3	4
Number of blocks (B)				

ii

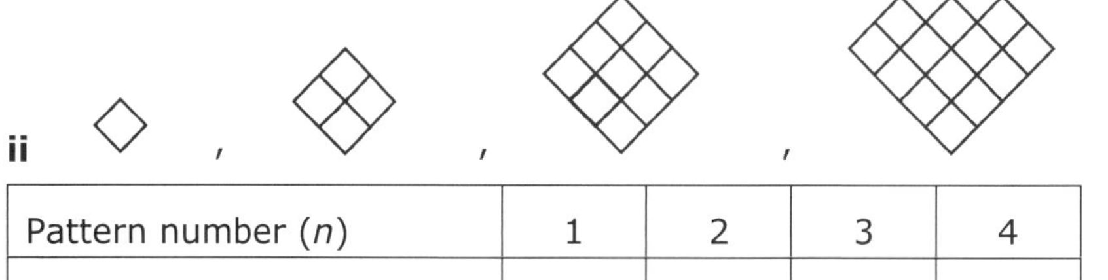

Pattern number (n)	1	2	3	4
Number of rhombuses (R)				

b In each case, find a formula connecting the given quantities.

c Explain, in each case, **why** the formula works.

4 Mrs Lovely has two children, who she spoils rotten. Every day she buys them presents.

On the first day, she buys them a present each. On the second day, she spoils them more so buys them two presents each. On the third day, three each and so on.

a How many presents will she have bought after 7 days?

b Write a formula connecting the number of presents and the number of days.

c Mrs Lovely began buying presents on 1st March. Use your formula to find out the number of presents she had bought by the end of the month.

d On what date will the number of presents exceed 5000?

Level 6

a Look at these numbers.

1^6 2^5 3^4 4^3 5^2 6^1

Which is the **largest**? *1 mark*

Which is equal to $\mathbf{9^2}$? *1 mark*

b Which **two** of the numbers below are **not** square numbers?

2^4 2^5 2^6 2^7 2^8

1 mark

Level 7

Each term of a number sequence is made by adding 1 to the numerator and 2 to the denominator of the previous term.

Here is the beginning of the number sequence:

$\frac{1}{3}, \quad \frac{2}{5}, \quad \frac{3}{7}, \quad \frac{4}{9}, \quad \frac{5}{11}, \quad \ldots$

a Write an expression for the ***n*th term** of the sequence. *1 mark*

b The *n*th term of a different sequence is $\frac{n}{n^2+1}$

The **first term** of the sequence is $\frac{1}{2}$.

Write down the **next three** terms. *2 marks*

S1.1HW Parallel lines

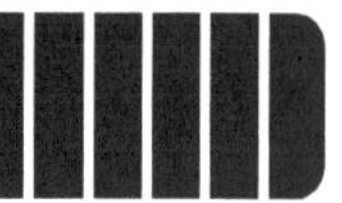

Remember:
- Corresponding angles are equal.
- Alternate angles are equal.
- Vertically opposite angles are equal.
- Supplementary angles add up to 180°.

Calculate the unknown angles, giving reasons for your answers.

1

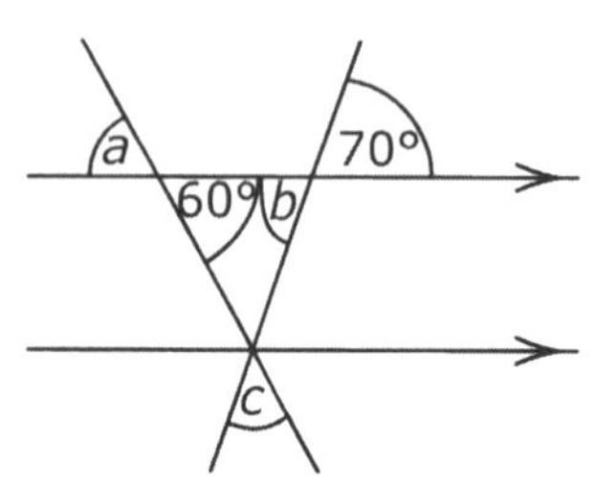

2

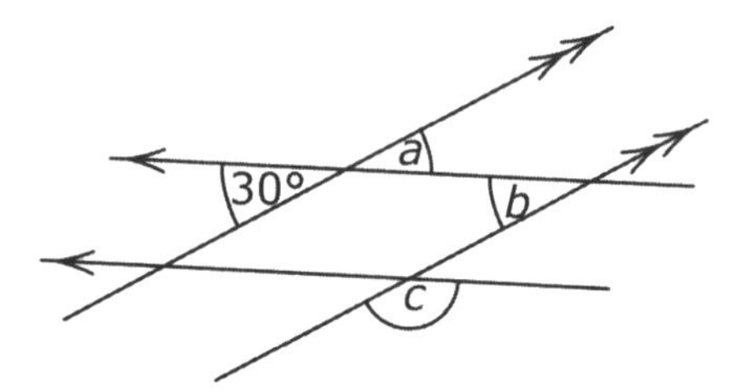

3

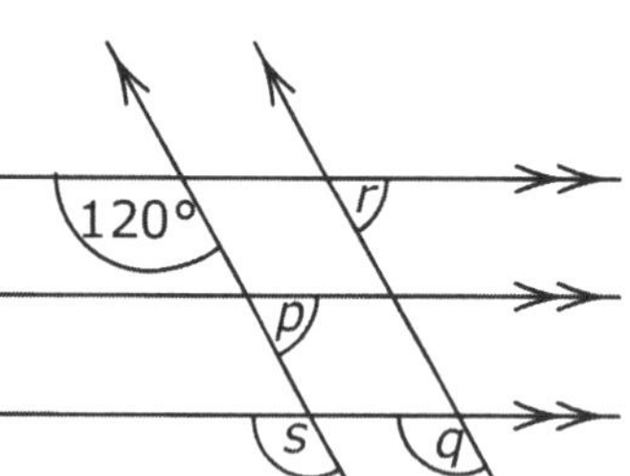

4

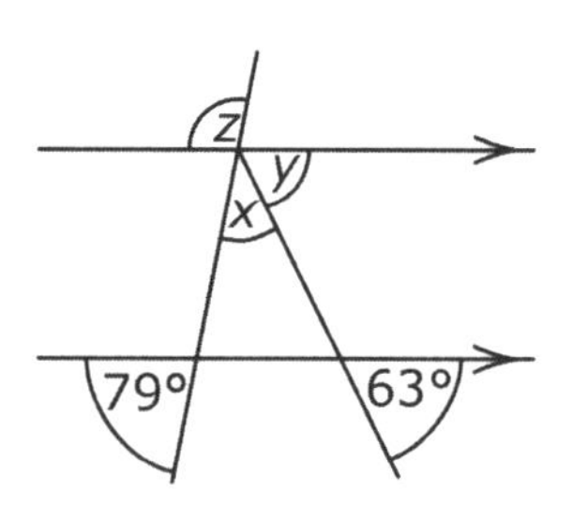

5

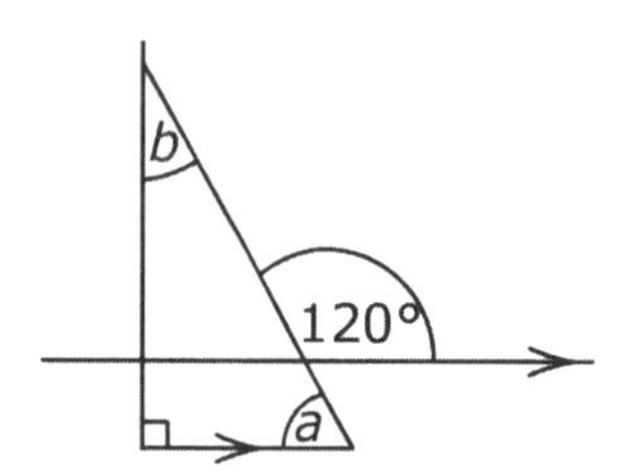

6

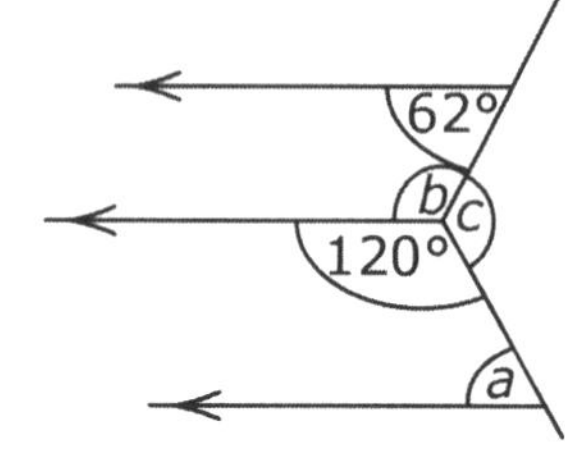

7

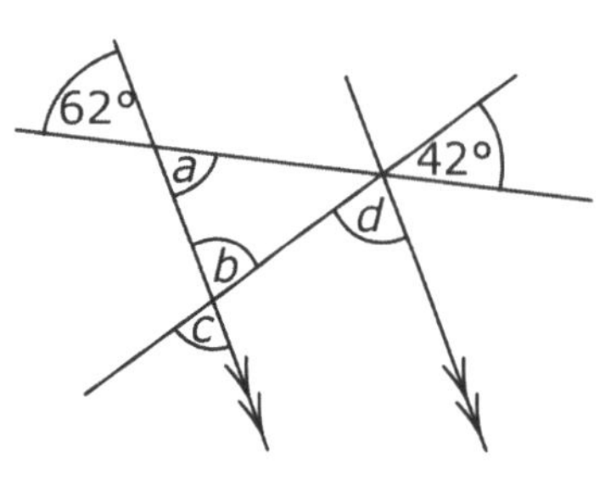

8

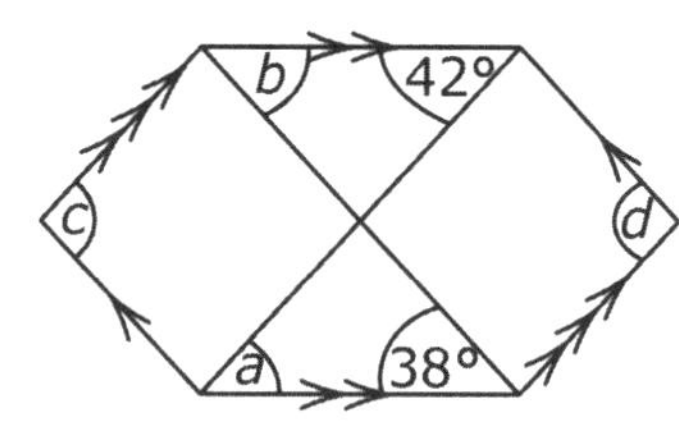

9

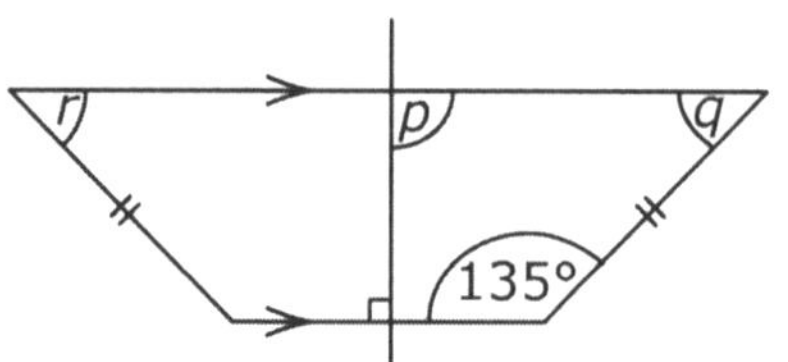

10

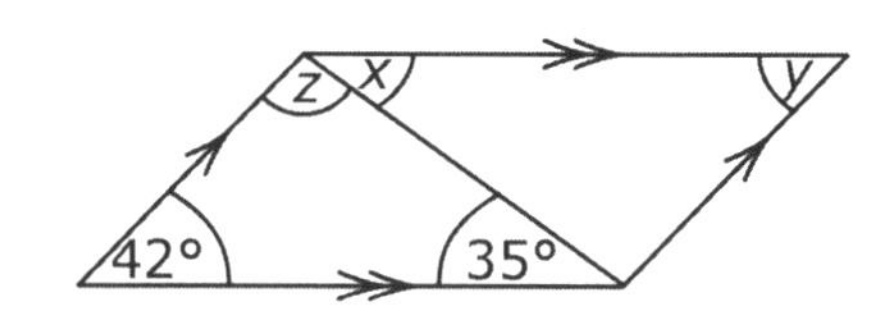

S1.2HW Triangle properties

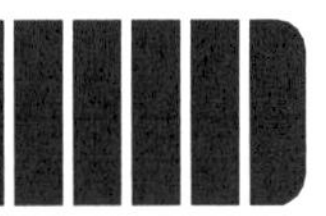

Remember:

◆ The exterior angle of a triangle is equal to the sum of the two interior opposite angles.

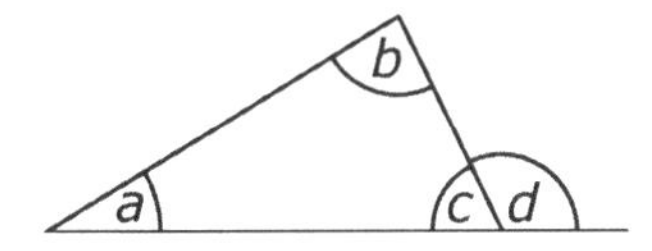

$c + d = 180°$

$a + b = d$

Find the unknown angles in these triangle diagrams.

1

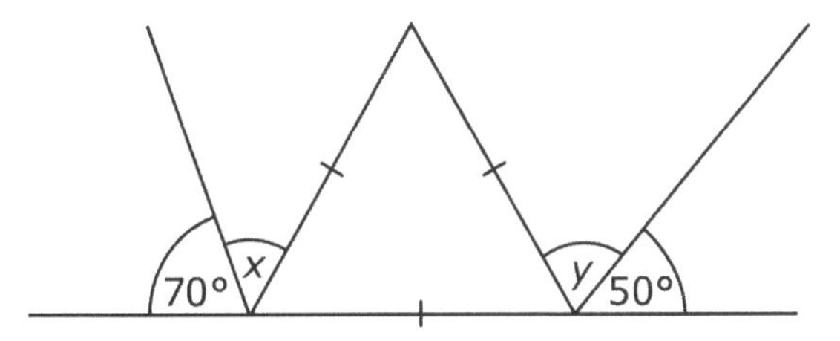

2

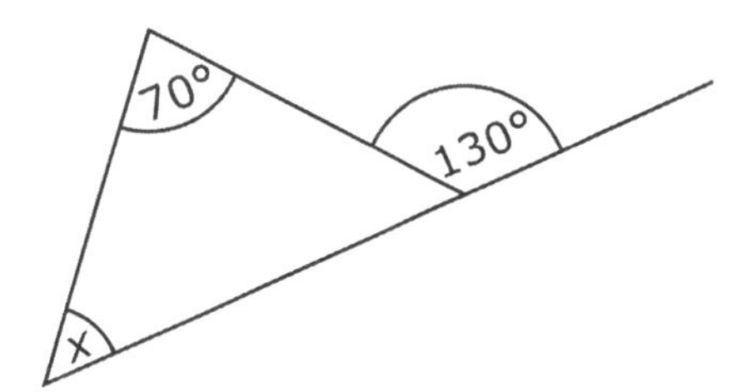

3

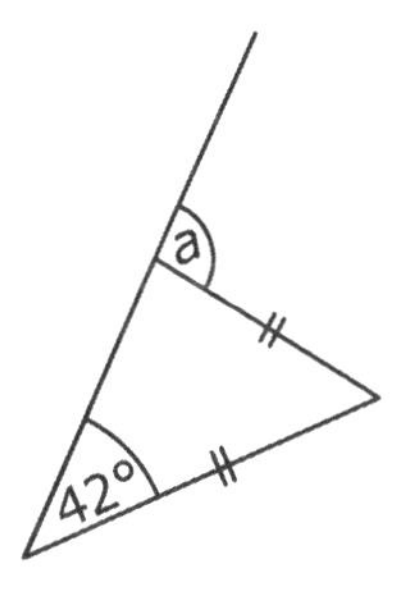

4

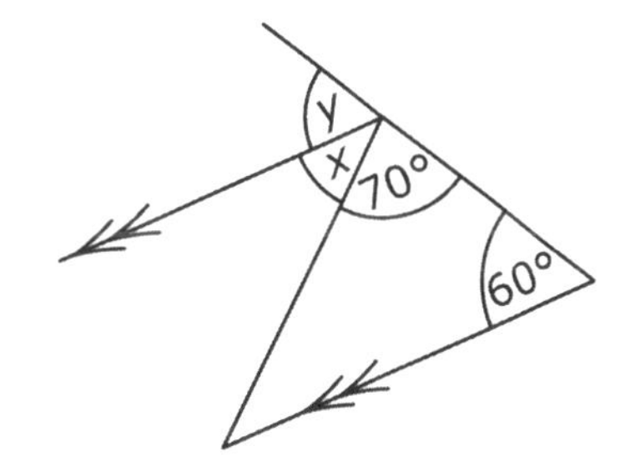

5

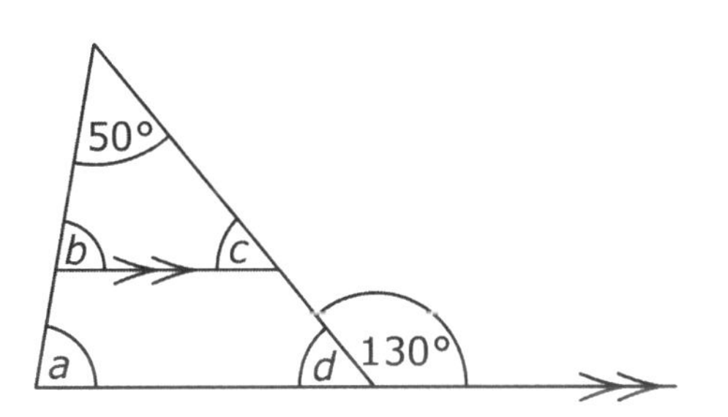

6

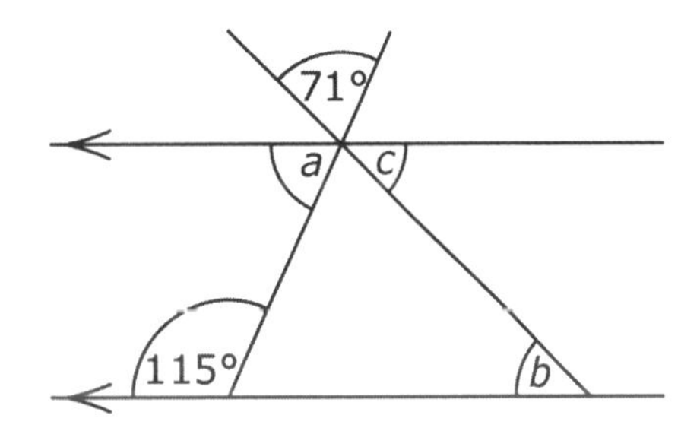

S1.3HW Calculating angles

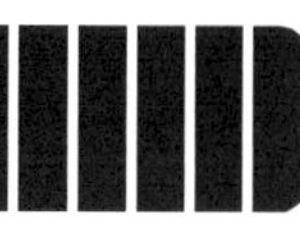

Remember:

- The interior angle sum S of any polygon with n sides is given by the formula:
  $S = (n - 2) \times 180°$
- The sum of the exterior angles of any polygon is 360°
- The interior angle of a regular polygon is $S \div n$ where S is the interior angle sum and n is the number of sides.

1 Calculate the unknown angles in these diagrams.

a

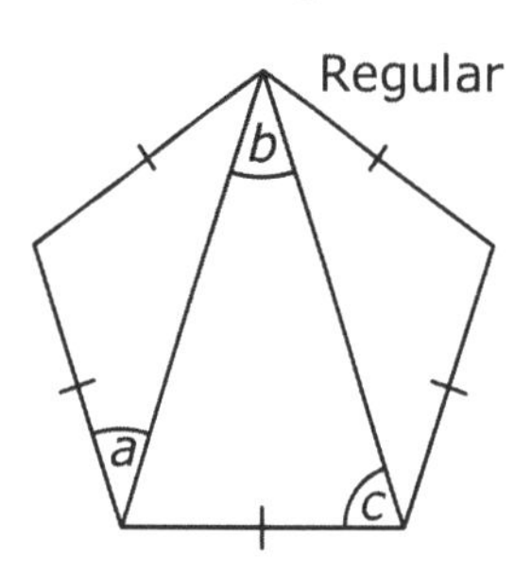

b

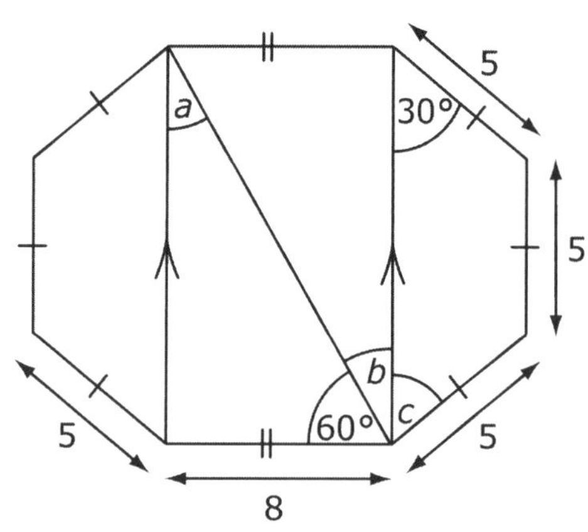

c

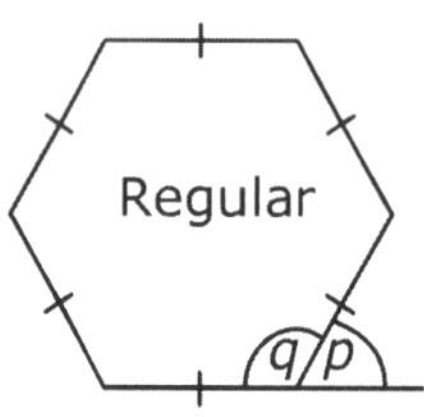

d

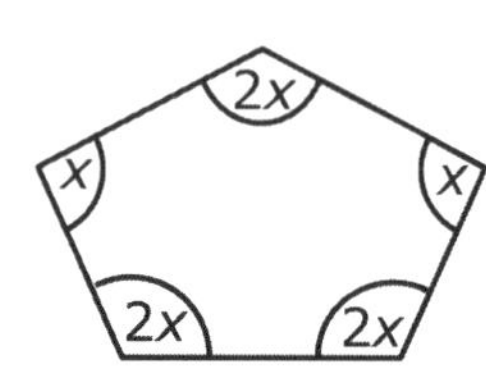

d

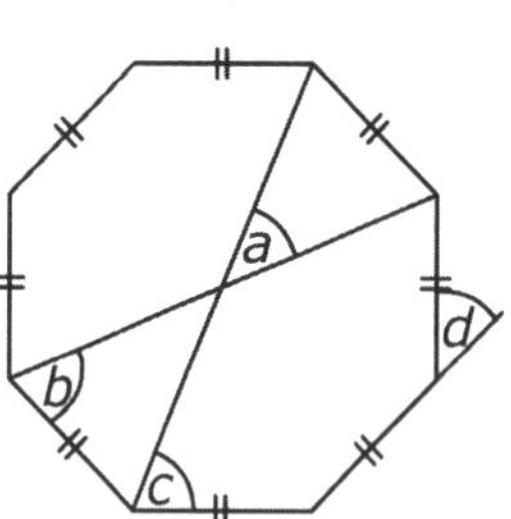

e

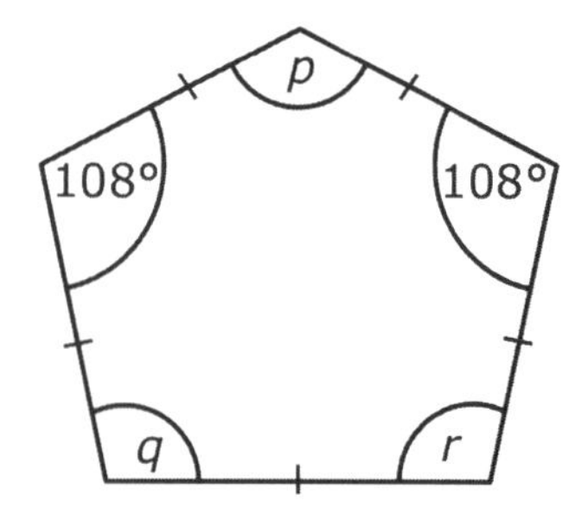

2 How many sides does a regular polygon have if it has interior angles of:

a 150° **b** 120° **c** 170°?

In each case, explain your answer.

3 Is it possible to have a regular polygon with interior angles of 112°? Explain your answer.

S1.4HW Circle properties

Remember:

- To bisect an angle:

1 Draw an angle XAY.

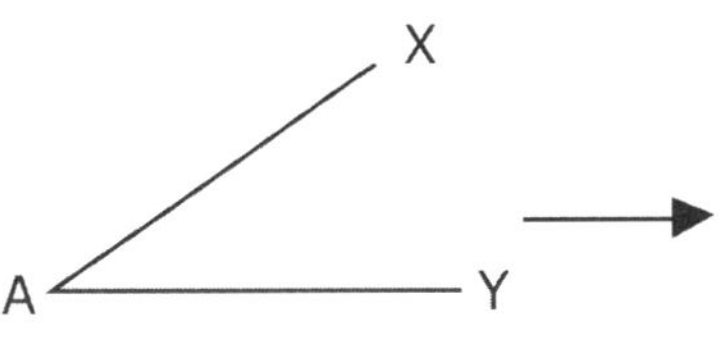

2 Open your compasses, put the point at A and draw an arc that cuts both lines.

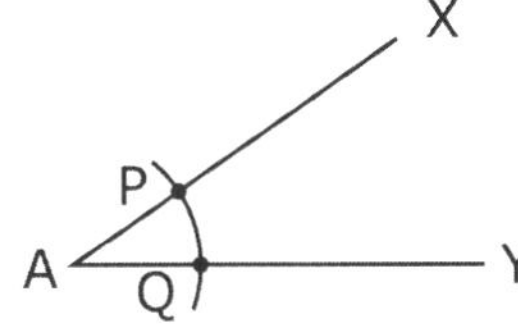

3 Draw arcs from P and Q using the same compass settings for each one.

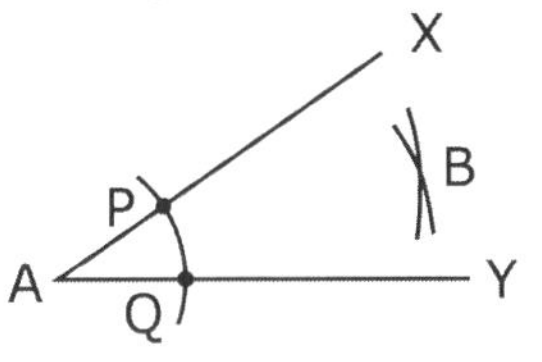

4 Join AB. This bisects the angle XAY

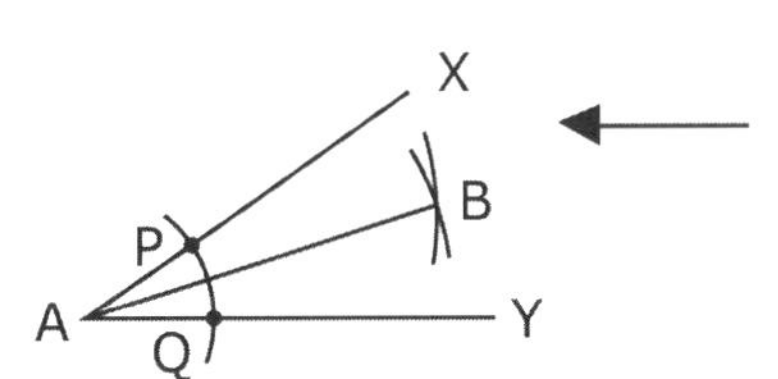

1 Draw an equilateral triangle of side 9 cm.
Use constructions to bisect each angle.

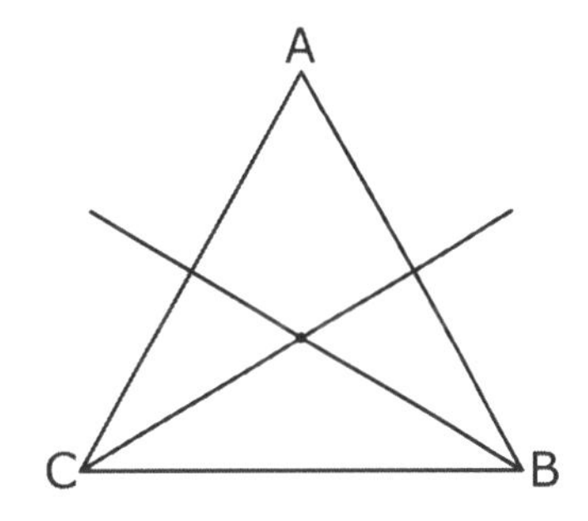

a What do you notice about the lines which bisect each angle?

b Label the point at the centre P.
Draw a circle touching side AB with centre P.

What do you notice about this circle?

You have constructed an **inscribed circle**.

2 Calculate the angles in the following circles.

a

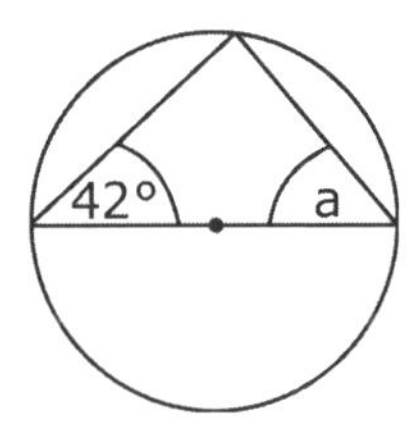

b

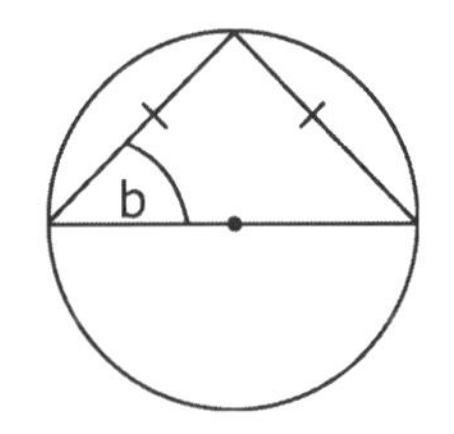

c

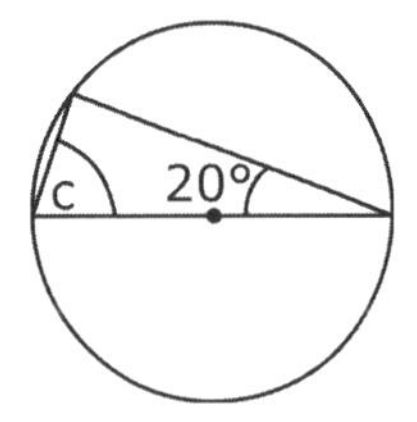

d

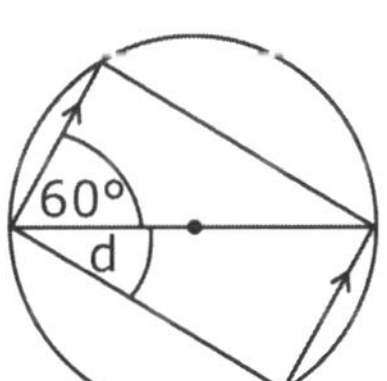

e

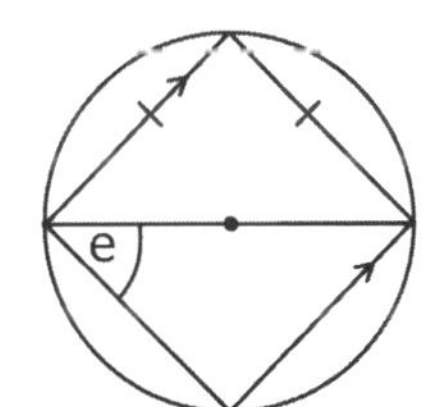

f

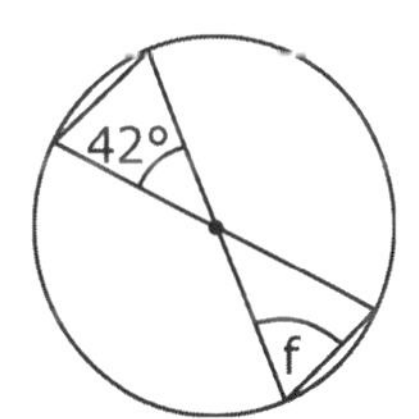

S1.5HW Constructing triangles

1 **a** Construct triangle ΔPQR where $PQ = QR = RP = 10$ cm.

b Bisect each side and join the points of bisection to form another triangle ΔXYZ. What do you notice about XY and PR?

c Bisect XY, YZ and XZ to form another triangle ΔABC. What do you notice about AC, XY and PR?

d Draw the next triangle and predict the result for the next side.

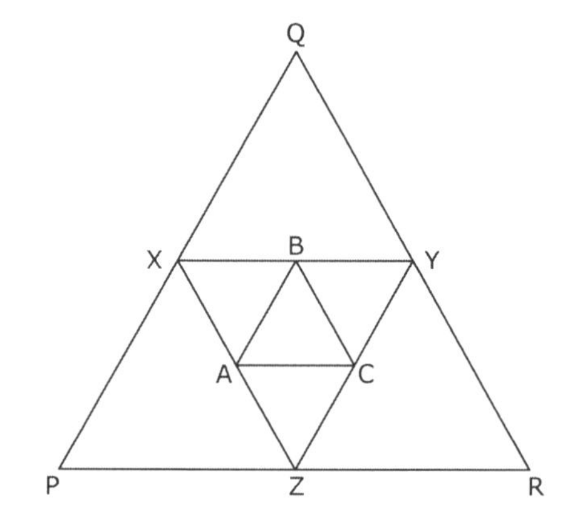

2 Construct ΔPQR with $PQ = 6$ cm, $QR = 8$ cm, $RP = 7$ cm.

3 Construct ΔXYZ with $XY = 5$ cm, $\hat{X} = 40°$, $\hat{Y} = 50°$.

4 Construct ΔABC with $AB = 6.2$ cm, $BC = 7.4$ cm, $\hat{B} = 43°$

5 Using only a pair of compasses construct:

a ΔXYZ where $XY = YZ = XZ = 5.4$ cm.

b ΔPQR, $PQ = QR = 4.2$ cm, $\hat{Q} = 60°$.

6 Construct ΔDEF where $DE = 8$ cm, $EF = 9$ cm, $DF = 7$ cm.
Construct the perpendicular bisector of each side.
Label the point where the three bisectors meet P.
Draw a circle centre P that passes through points D, E and F.

Constructing right-angled triangles

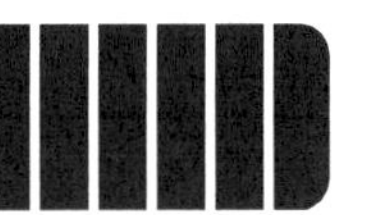

1 Construct $\triangle$ABC where $\hat{A} = 90°$, BC = 7.8 cm, AC = 5.7 cm.
Measure AB.

2 Construct $\triangle$PQR where $\hat{P} = 90°$, PQ = 10 cm, PR = 8 cm.
Measure QR.

3 Construct $\triangle$ABC where AB = 5 cm, BC = 12 cm, CA = 13 cm.

4 Construct $\triangle$EFG where $\hat{P} = 90°$, EF = FG = 8 cm.

5 Construct $\triangle$XYZ where XY = 9 cm, $\hat{X} = 60°$, $\hat{Y} = 70°$.

6 A farmer has a triangular vegetable patch as shown.

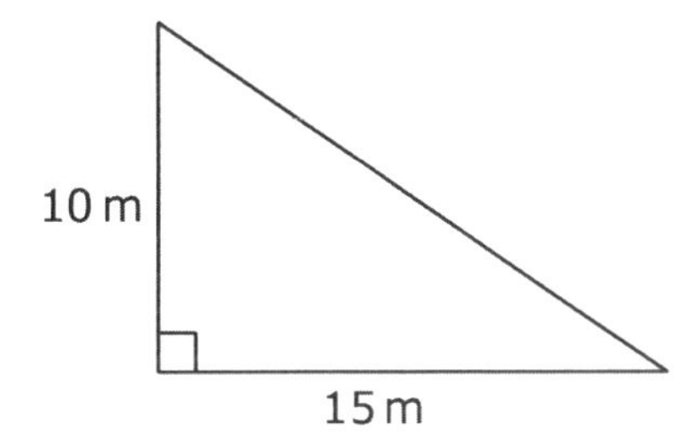

Draw a scale diagram of the vegetable patch using a scale of 1 cm to 1 m.
Measure the length of the hypotenuse.
Hence find the amount of fencing needed to enclose the vegetable patch completely (no gates required).

Level 6

The diagram shows two isosceles triangles inside a parallelogram.

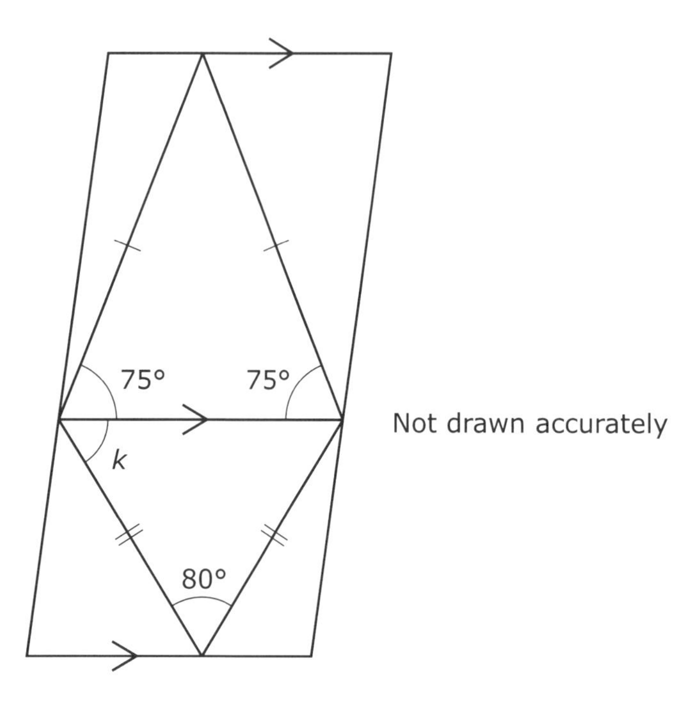

Not drawn accurately

a Copy the diagram. On your diagram, mark another angle that is 75°.

Label it 75°. *1 mark*

b Calculate the size of the angle marked k.

Show your working. *2 marks*

Level 7

The diagram shows a **rectangle** that just touches an **equilateral triangle.**

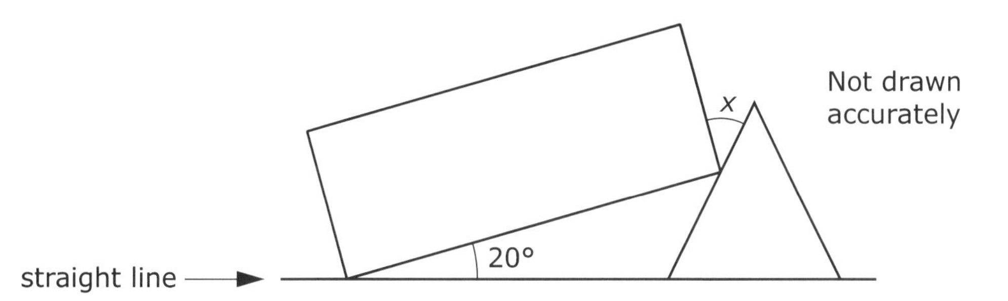

a Find the size of the angle marked x.
Show your working. *2 marks*

b Now the rectangle just touches the equilateral triangle so that **ABC** is a **straight line.**

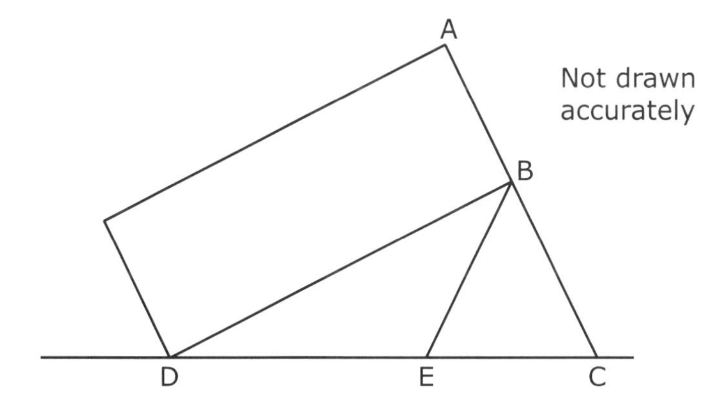

Show that **triangle BDE** is **isosceles**. *2 marks*

D1.1HW Theoretical probability

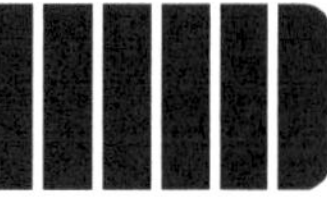

Remember:

◆ Theoretical probability of an event = $\dfrac{\text{number of favourable outcomes}}{\text{total number of outcomes in sample space}}$

1 At the 'Hook-A-Duck' stall at a school fete, children use a rod with a hook at the end to catch a duck floating in a large bowl of water.

The underside of each duck is marked either S for star prize, W for win or L for lose. After a duck has been hooked it is always returned to the water before another duck is hooked.

a For the trial 'hooking-a-duck', list the outcomes in the sample space.

b For the trial described in **a** discuss whether or not you think the outcomes will be equally likely.

In the bowl there is 1 duck marked S, 9 marked W and 2 marked L.

c Write down the theoretical probability of the following events.

i Winning the star prize
ii Not winning the star prize
iii Winning something
iv Not winning any prize
v Just a win

2 Another stall at the school fete has this board:

				PRIZE!	
1	2	3	4	5	6

You throw a dice.

You win a prize if your score is 5.

You win a penny sweet if your score is an even number.

a List the outcomes in the sample space.

b Write down the theoretical probability of the following events.

i winning something
ii not winning anything
iii not winning the prize
iv not winning a penny sweet

D1.2HW Recording outcomes

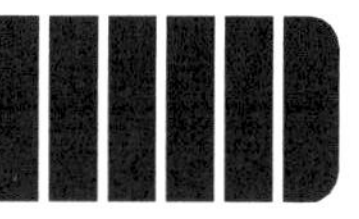

Karl has three fair tetrahedral dice X, Y and Z.

The numbers on dice X are 1, 5, 5, 1

The numbers on dice Y are 4, 4, 4, 0

The numbers on dice Z are 2, 2, 6, 2

Dice X

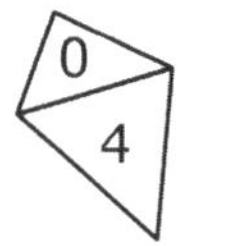

Dice Y

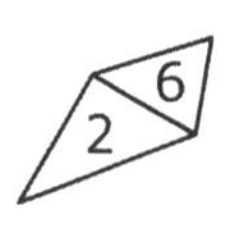

Dice Z

Tetrahedral dice have four sides.

Two dice are chosen and rolled at the same time.

The dice that shows the highest score is the winning dice.

1 Copy and complete the sample space diagram to show which dice wins when X and Y are rolled together.

Dice Y

		4	4	4	0
Dice X	1	Y			
	5			X	
	5		X		
	1	Y			

What is the probability that X wins against Y?

2 Draw a sample space diagram to show which dice wins when Y and Z are rolled together. What is the probability that Y wins against Z?

3 Draw a sample space diagram to show which dice wins when Z and X are rolled together. What is the probability that Z wins against X?

4 Comment on the probability of winning with each dice.

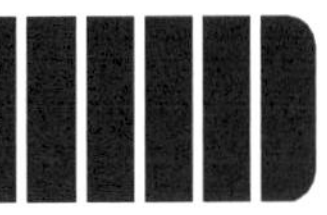

D1.3HW More theoretical probability

Three fair tetrahedral dice X, Y and Z are thrown at the same time.

The numbers on dice X are 1, 5, 5, 1

The numbers on dice Y are 4, 4, 4, 0

The numbers on dice Z are 2, 2, 6, 2

Dice X

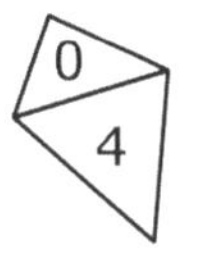

Dice Y

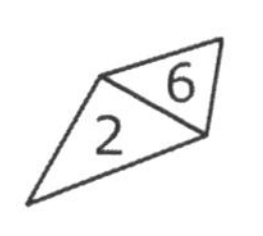

Dice Z

Tetrahedral dice have four sides.

1 Copy and complete the tree diagram to show the different probabilities of the possible outcomes when these three dice are thrown.

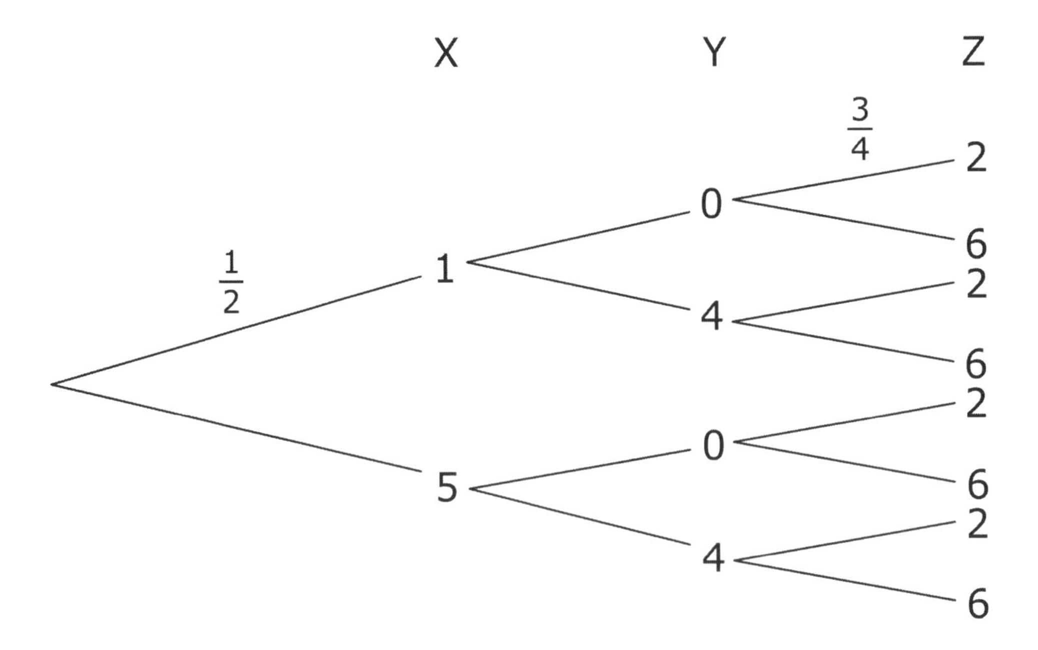

2 The scores shown on each dice are added together.

a What is the highest possible score?

b Work out the probability of getting the highest possible total score.

D1.4HW Experimental probability

Remember:

- Estimated probability = $\frac{\text{number of successes}}{\text{total number of trials}}$

Make a dice out of card.

You can make an ordinary six-sided dice or a tetrahedral (four-sided) dice.

Use the nets below to help with the construction.

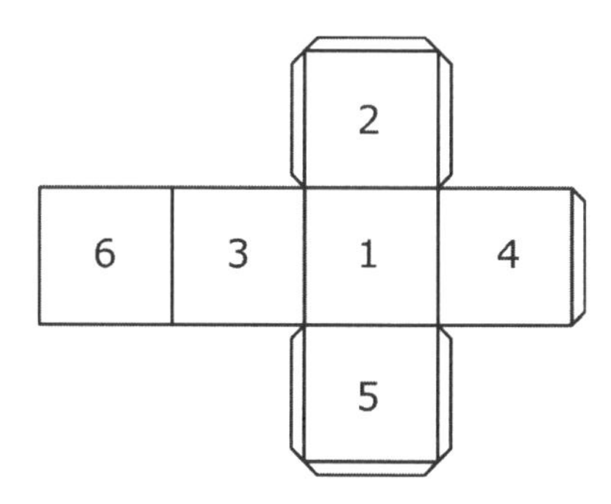

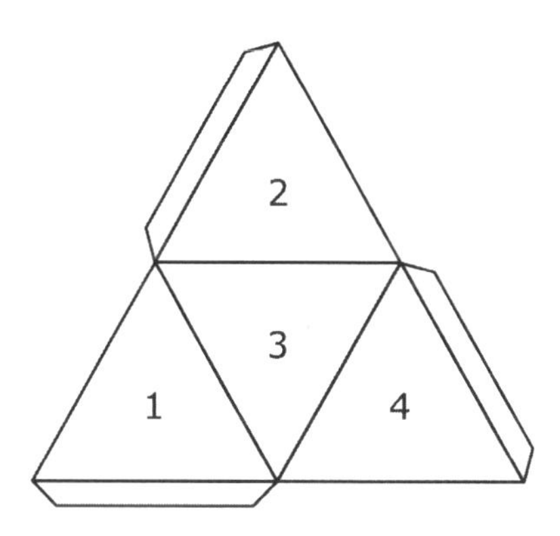

Weight the inside of one of the faces on your dice with a small amount of Blu-tack or sellotape.

Carry out an experiment to estimate the probability of throwing each number on your dice.

Dice game

Design a game for two or more people to play with one or more dice.

The dice can be six-sided or tetrahedral.

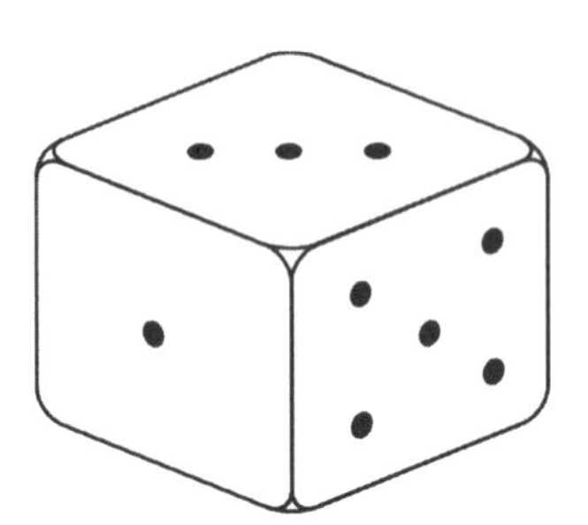

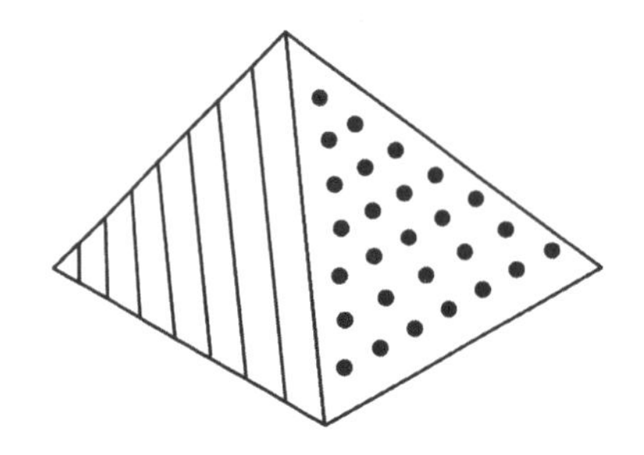

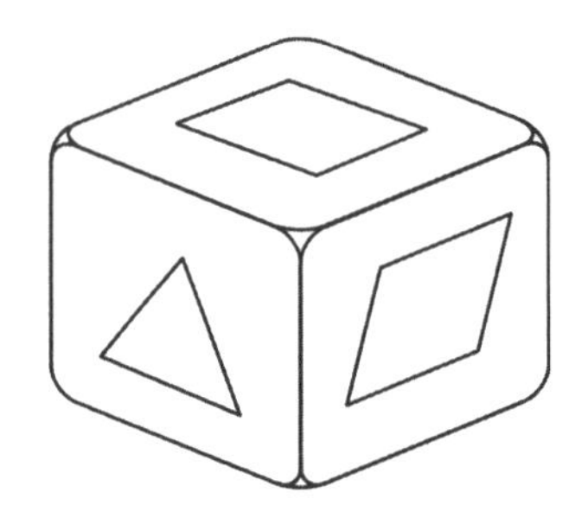

The dice can be numbered or have patterned faces or faces showing shapes.

Write out a set of rules to play your game.

Play your game to see how long it takes before you have a winner.

Here is a net for a tetrahedral dice:

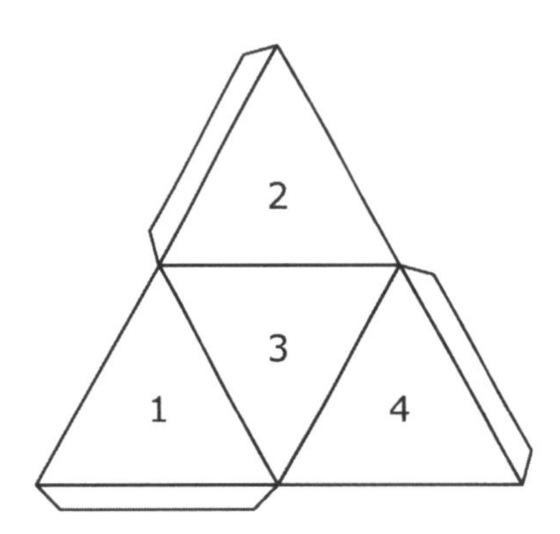

D1.6HW Comparing probabilities

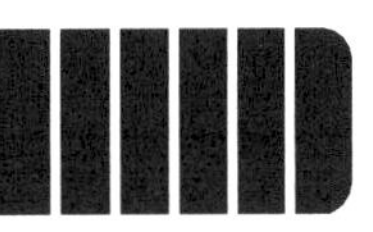

A tetrahedral dice has four faces, coloured orange, red, green and yellow. The dice is biased to the colour yellow, so that the probability of yellow when the dice is thrown is $\frac{1}{2}$.

The probability of getting orange, red and green is $\frac{1}{6}$ for each colour.

You can use the extract of random number tables below to simulate the colour you would get when you throw the dice.

Let 1 be orange, 2 be red, 3 be green and 4, 5 and 6 be yellow.

Ignore the numbers 7, 8, 9 and 0.

49487 52802 28667 62058 87822 14704

29480 91539 46317 84803 86056 62812

25252 97738 23901 11106 86864 55808

02341 42193 96960 19620 29188 05863

69414 89353 70724 67893 23218 72452

1 Start at the second row and use the simulation to write down the colours you would get from 20 throws of this dice.

The first digits are 294 → red, (ignore 9), yellow

Use these results to write down the probability of getting each colour.

2 Choose a different starting point and use simulation to write down the colours you would get from 20 throws of this dice.
Use these results to write down the probability of getting each colour.

3 Compare the theoretical probability of getting each colour with the experimental probabilities you found from simulation.
Comment on your comparisons.

Level 6

I have two bags of counters.

Bag A contains
12 red counters and
18 yellow counters.

Bag B contains
10 red counters and
16 yellow counters.

I am going to take one counter at random from either bag A or bag B

I want to get a **red** counter.
Which bag should I choose?

Show working to explain your answer. *2 marks*

Level 7

Some pupils threw 3 fair dice.

They recorded how many times the numbers on the dice were the same.

Name	Number of throws	Results		
		all different	2 the same	all the same
Morgan	40	26	12	2
Sue	140	81	56	3
Zenta	20	10	10	0
Ali	100	54	42	4

a Write the name of the pupil whose data are **most likely** to give the best estimate of the probability of getting each result.

Explain your answer. *1 mark*

b This table shows the pupils' results collected together:

Number of throws	Results		
	all different	2 the same	all the same
300	171	120	9

Use these data to estimate the **probability** of throwing numbers that are **all different**. *1 mark*

c The theoretical probability of each result is shown below:

	all different	2 the same	all the same
Probability	$\frac{5}{9}$	$\frac{5}{12}$	$\frac{1}{36}$

Use these probabilities to calculate, for 300 throws, **how many times** you would theoretically expect to get each result. *2 marks*

d Explain why the pupils' results are not the same as the theoretical results. *1 mark*

e Jenny throws the 3 dice twice.

Calculate the probability that she gets **all the same** on her first throw and gets **all the same** on her second throw.

Show your working. *2 marks*

Fractions and decimals

1 The number of students in each year of Ivywell School is given in the table.

Year	Number of students	Proportion
7	120	
8	130	
9	134	
10	144	
11	112	

a Copy the table. Insert the proportions shown in Box A in the correct place in the table.

b Lucia says you should write all your proportions as fractions.
Alex disagrees. Explain why Alex might disagree.

Box A

$\frac{9}{40}$ $\frac{3}{16}$ 0.209375

$\frac{7}{40}$ 0.203125

2 The diagram shows a rectangle ABCD.

a What fraction is triangle OFG of:
i $\triangle$FGH **ii** $\triangle$ADO?

b Copy the diagram and show one way in which you could shade $\frac{1}{64}$ of the rectangle ABCD.

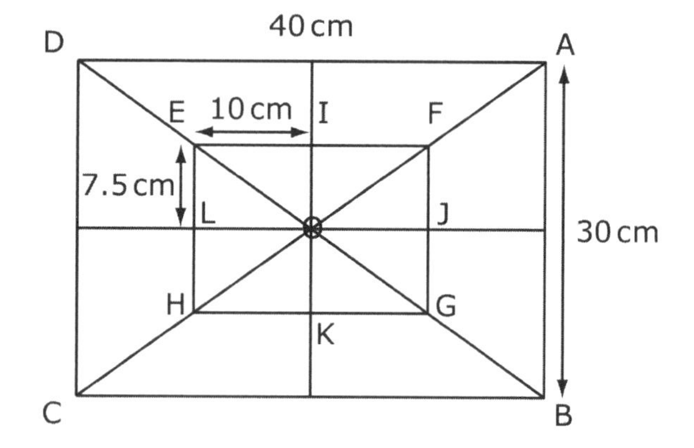

N2.2HW Adding and subtracting fractions

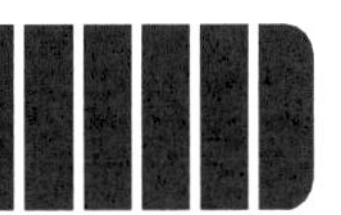

1 Look at these fractions.

$$\frac{34}{48} \qquad \frac{23}{25} \qquad \frac{14}{81}$$

a Arrange them in order smallest to largest.

For the following calculations, express each answer in its simplest form.

b Use the 'fraction' method – do not convert to decimals – to subtract the smallest from the largest.

c Add all three fractions together.

d What fraction lies exactly halfway between the smallest and the next largest fraction?

2 Work out these fraction sums.
Use a fraction method – do not change them to decimals.
Show your working clearly, and write your answers as mixed numbers of fractions in their simplest form.

a $\frac{11}{30} + \frac{3}{16} + \frac{2}{18}$

b $\frac{9}{27} - \frac{1}{6} - \frac{1}{24}$

c $12\frac{1}{5} + \frac{3}{10} - 6\frac{11}{15}$

d $3\frac{3}{7} - 1\frac{5}{7} + 8\frac{9}{11}$

e $\frac{10}{3} + \frac{1}{29} - \frac{1}{16} + \frac{3}{9}$

N2.3HW Fractions of amounts

1 The rectangle below has side lengths a and b.

b

a Area

Work out each of the following:

a What is a if Area $= 45\frac{1}{3}$ cm^2 and $b = 6\frac{2}{3}$ cm ?

b What is b if Area $= 45\frac{1}{3}$ cm^2 and $a = 5\frac{1}{8}$ cm ?

c What is Area if a $= 7\frac{1}{4}$ cm^2 and $b = 8\frac{4}{7}$ cm ?

d What is a if Area $= 9\frac{3}{4}$ cm^2 and $b = 3\frac{7}{8}$ cm ?

e What is a if Area $= 6\frac{2}{5}$ cm^2 and $b = 3\frac{7}{10}$ cm ?

2 **Challenge**

In a game, contestants all have a share of a jar containing 1 000 sweets. The winning contestant from each round walks away with $\frac{1}{3}$ of the remaining sweets. In round 1, the winner receives 333 sweets (the number of winner's sweets is always rounded down). So, the remaining contestants have 667 sweets to play for!

a How many sweets will the winner of round 2 receive?

b How many sweets are left **in total** for round 3?

c How many sweets will the winner of round 3 receive?

d How many sweets will the winner of round 4 receive?

e For how many rounds can the game continue before the winner has no sweets?

N2.4HW Percentages of amounts

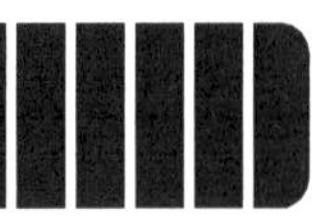

Fred's marks for each of his 'end of year' tests in his school subjects were:

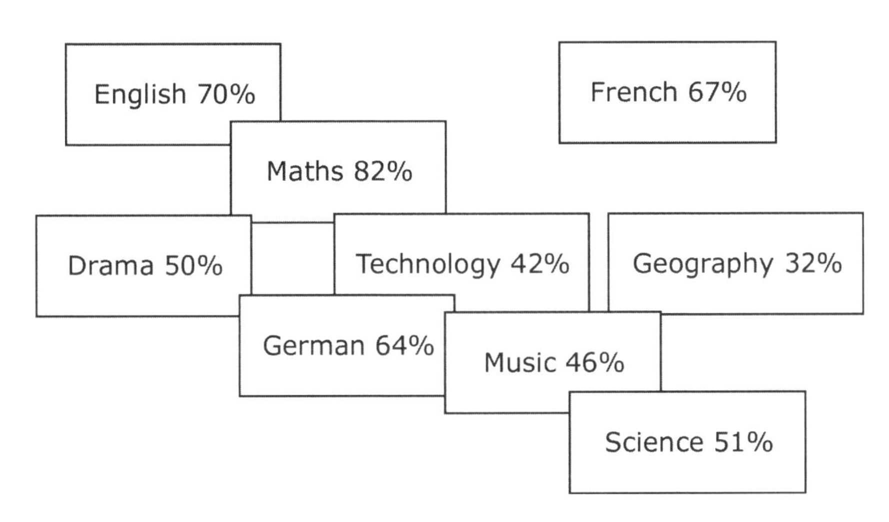

He set himself the following targets:

- the subjects in which he scored 50% or less he would like to improve by 14%
- the subjects in which he scored more than 50% he would like to improve by 7%.

Work out his target percentage for each subject, to the nearest whole number.

For example, for a score of 38%:

14% of 38 = 5.32

His new target is 38 + 5.32 = 43.32

= 43% to the nearest whole number.

Explain a better way that he could set his targets.

N2.5HW Problems involving percentage change

1 Work out the original price of each of the following items.

a

Jeans
Reduced by 18%
Now £20

b

Trainers
Reduced by 34%
Now £60

c

Coats
Reduced by 18%
Now £41

d

Scarves
Reduced by 48%
Now £14

2 Andy, Sara and Elaine all have credit cards.

a Andy owes £3000 on his card.
The bank adds on interest at 5% per month.
Andy pays £400 back to the bank per month.
How much does he still owe after six months?

b Sara owes £5000.
The interest rate for her card is 10% per month, but she pays back £500 per month.
How much will she owe after six months?

c Elaine owes £2000.
The bank adds on 6% interest per month, and she pays back £200 per month.
How much will she owe after six months?

d Who will pay off their credit card bill first – Andy, Sara or Elaine?

Comparing proportion

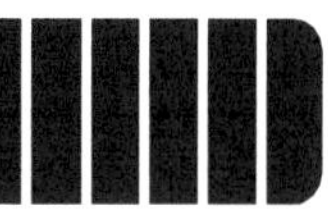

nvestigation

n an abrasives factory the cost of one piece of sandpaper is 0.583471p.
he sandpaper is put into packs of 20.
hese packs are put into boxes of 50.
he boxes are sent out of the factory in pallets of 220.

he table shows the charge for a unit, a pack, a box and for a pallet.

Unit (1 piece of sandpaper)	0.583471p
Pack (= 20 units)	11.67p
Box (= 50 packs)	£5.84
Pallet (= 220 boxes)	£1285

a A customer orders 1.76 million units of sandpaper.

i If the customer is charged per unit, how much is the bill?

ii If the customer is charged per pallet how much is the bill? Why is there a difference in the two bills?

iii Investigate the price for being charged per packs and boxes.

b Investigate how the abrasives company have arrived at all their prices for packs, boxes and pallets.

c i What is the percentage difference in the prices for 1 million units sold as units, packs, boxes and pallets.

ii Which is the most cost-effective way to buy sandpaper?

Level 6

$\frac{1}{3}, \frac{1}{8}, \frac{1}{5}$ are all examples of unit fractions.

All unit fractions must have

$\frac{1}{3}$ — a numerator that is 1

— a denominator that is an integer greater than 1

The ancient Egyptians used only unit fractions.

For $\frac{3}{4}$, they wrote the sum $\frac{1}{2} + \frac{1}{4}$.

a For what fraction did they write the sum $\frac{1}{2} + \frac{1}{5}$?

Show your working. *1 mark*

b They wrote $\frac{9}{20}$ as the sum of two unit fractions.

One of them was $\frac{1}{4}$.

What was the other? *1 mark*

Show your working *1 mark*

Level 7

a One calculation below gives the answer to the question:

What is 70 increased by 9%?

Write down the correct one.

70×0.9	70×1.9	70×0.09	70×1.09

1 mark

Choose one of the other calculations.

Write a question **about percentages** that this calculation represents. *1 mark*

Now do the same for one of the remaining two calculations.

1 mark

b Copy and complete this statement with the missing decimal number.

To decrease by 14%, multiply by ______________ *1 mark*

A2.1HW Indices

1 Copy and complete the crossword, without using a calculator:

1			2
3		4	
		5	6
	7		

Across

1 $4^2 \div 2^3$

3 $12^2 + 13^2 + 14^2$

5 $9^2 + (-3)^2 + 1$

7 $4 \times (\frac{1}{2})^2$

Down

1 8

2 1^{16}

4 1 less than the 10th square number

6 $12^2 \div 8$

2 Put the following cards in ascending order:

0^{20} $(-2)^5$ 1^{50} $(\frac{1}{4})^3$ 0.2^3

3 Simplify these expressions. Give your answer in index form.

a $p^2 \times p^3$ **b** $q^6 \times q^8$ **c** $m^4 \times m^{-2}$

d $a^7 \div a^4$ **e** $b^{11} \div b^{14}$ **f** $\dfrac{c^{10}}{c^{-2}}$

g $(k^4)^2$ **h** $(n^{11})^{-2}$ **i** $(2m^2)^4$

4 Copy and complete:

a $m^7 \times m^{\square} = m^{15}$ **b** $q^{\square} \div q^{11} = q^8$

c $(p^3)^{\square} = p^{24}$ **d** $z^{\square} \times z^8 = z^3$

5

$5^8 = 390\,625$
$5^9 = 1\,953\,125$
$5^{10} = 9\,765\,625$
$5^{11} = 48\,828\,125$

Use the information given to work out:

a $5^3 \times 5^6$

b $(5^2)^4$

c $(5^6)^4 \div (5^3 \times 5^{10})$

d 5^7

6 Simplify these expressions. Give your answers in index form.

a $x^7 (x^6 + x^4)$ **b** $x^{-3} (x^{-2} + x^9)$ **c** $\dfrac{y^{10} + y^8}{y^2}$

A2.2HW Further indices

1 Match pairs of expressions with equal values:

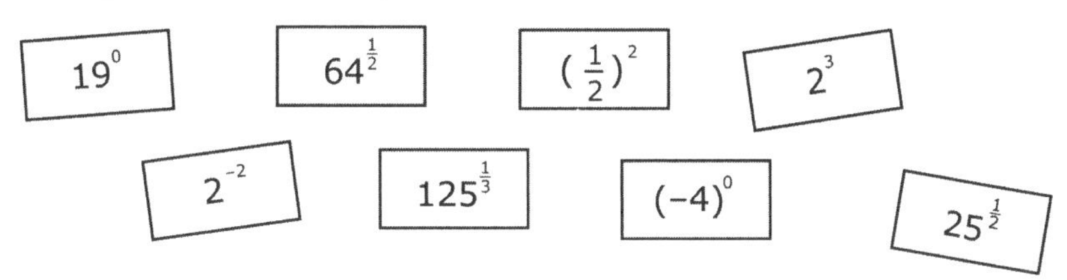

2 Evaluate these indices. Use the rules of indices first.

a $36^{\frac{1}{4}} \times 36^{\frac{1}{4}}$ **b** $5^3 \div 5^5$

c $(7^{-1})^2$ **d** $8^2 \times 8^{-2}$

3 Copy and complete:

a $25^{\square} = 5$ **b** $9^{\square} = \frac{1}{81}$

c $4^{\square} = 1$ **d** $64^{\square} = 4$

e $10^{\square} = \frac{1}{1\,000\,000}$ **f** $169^{\square} = 13$

g $2^{\square} = \frac{1}{8}$ **h** $1^{\square} = 1$

4 $64^m = 2^n = 8$

What are m and n?

5 **a** Explain why $(-36)^{\frac{1}{2}}$ is impossible to evaluate.

b Does $(-64)^{\frac{1}{3}}$ have a solution?

6 Use the following cards (you do not have to use them all) to write as many true index equations as you can.

eg. $8^0 = 1$

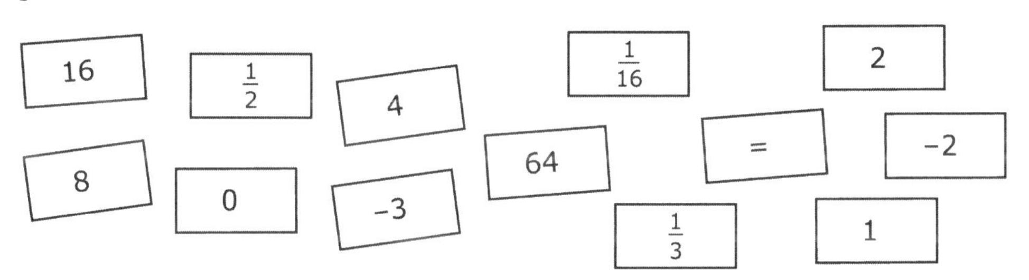

A2.3HW Simplifying expressions

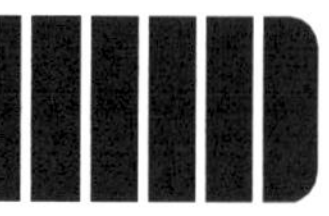

Copy the grid, replacing each expression with its simplified form.
(It may not be possible to simplify some expressions.)

$5a + 7a - 4a$	$3x \times 4y$	$9x + 3x^2 - 6x$	$p \times q \times q$
$6x \times 10x$	$5ab + 4ba$	$p \div q$	$36q \div 4$
$10x + 11x^2 + 12x^3$	$x^3 \times x^8$	$10x + 4y - 3x - 8y$	$(3p^2)^2$
$\frac{15a^2}{5a}$	$9t - 4s - 6t - 10s$	$\frac{45ab}{9b}$	$11 + x$
$11xy + 12xy - 3yx$	$5x^9 \times 6x^{10}$	$abcd + dcba$	$\frac{50x^{10}}{5x^3}$
$5x \times 3y \times 6x$	$-4x - 3x + 10x$	$x^2 \times x$	$6p + s - 4p^2$

A2.4HW Expanding brackets

1 Expand these expressions:

a $6(3x + 4)$ **b** $7(5 - 2b)$ **c** $q(q + 6)$

d $2p(3p + 4)$ **e** $-7(2x + 1)$ **f** $-3(2 - 5z)$

g $-p(p - q)$

2 Expand and simplify these expressions. Hence, show that they are all the same.

$3(x + 6) + 5(x + 2)$ $5(2x + 6) - 2(x + 1)$ $(10x - 2) - (2x - 30)$

$6(3 - 2x) - 10(-2x - 1)$

3 $5x + 10$ can be factorised to give $5(x + 2)$. This answer can be checked by expanding.
Copy and complete these factorisations:

a $6y + 12 = \square(y + 2)$ **b** $15z - 25 = \square(3z - 5)$

c $9p + 18 = \square(p + 2)$ **d** $16x + 24y + 40 = \square(2x + 3y + 5s)$

e $5w + w^2 = \square(5 + w)$ **f** $10 + 20x = 10(\square + \square x)$.

4 Write an expression for the required quantity. Simplify, then factorise your expression.

a

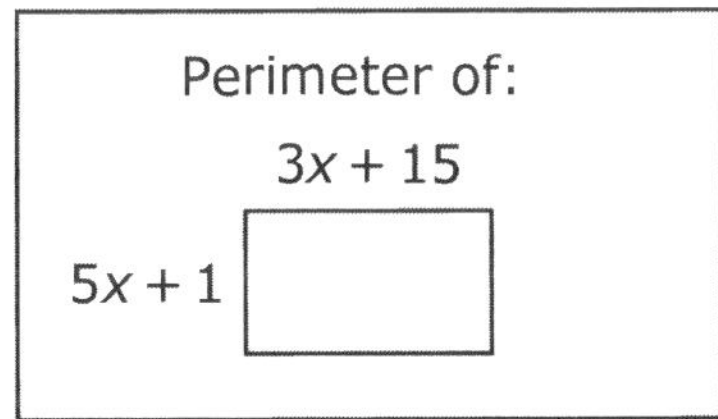

b

Sum of any four consecutive numbers:

$x, x + 1, \ldots$

c

Difference in area:

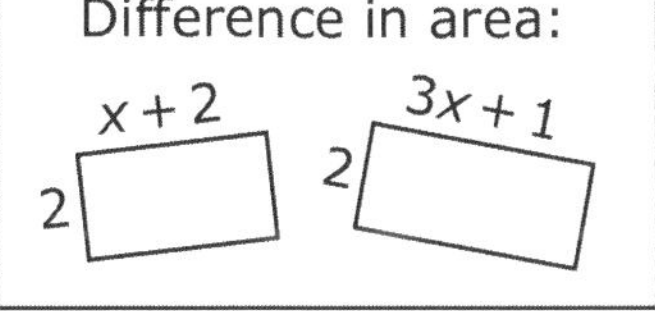

5 Correct these factorisations:

a $8x + 24 = 8(x + 16)$ **b** $3 + 9x = 3(0 + 3x)$

c $x^5 + x = x(5 + 1)$

Factorising

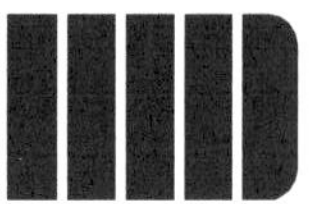

1 Factorise these expressions completely:

a $5x + 10$ **b** $6y - 12$ **c** $4x + 12y - 8z$

d $2a + a^2$ **e** $x^2 - x$ **f** $p^3 + p^2$

g $6ab + 3b$ **h** $10cd + 5de + 15d$

2 Explain why $13a + 37$ cannot be factorised.

3 Weird creatures have antennae on their heads.

a Copy and complete this table:

Number of heads h	1	2	3	4
Number of antennae a	4			

b Find a formula connecting the number of heads with the number of antennae.

c Explain why the formula works.

d Factorise your formula. Can you explain this new form?

4 Repeat question 3 for these sequences of diagrams:

a

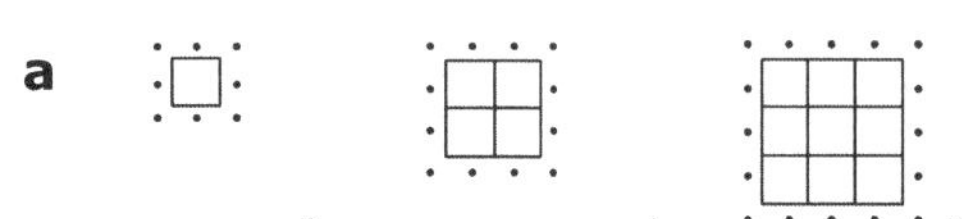

Length of side, ℓ	1	2	3	4
Number of dots, d	8			

b

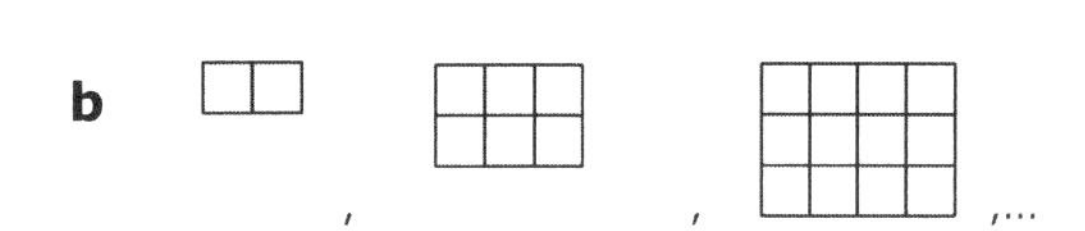

Length of side, ℓ	1	2	3
Number of tiles, t	2		

5 **a** Find an expression for the area of the kite by splitting it into triangles.

b Using factorisation, show that the area of a kite is "$\frac{1}{2}$ the product of the diagonals".

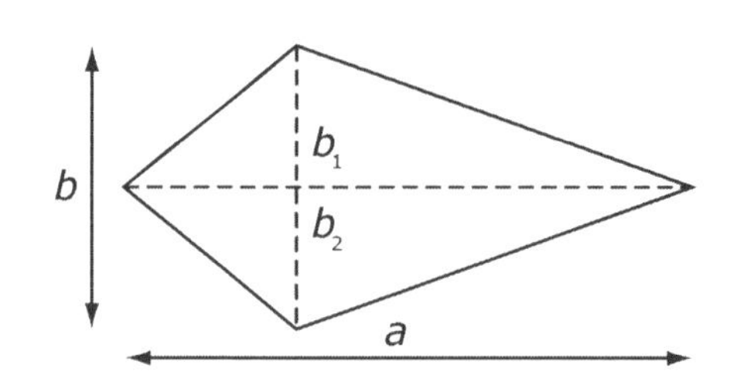

Remember:
Area of triangle
$= \frac{1}{2} \times$ base $\times$ height

A2.6HW Solving equations

1 Solve the following equations. Hence, show that they all have the same solution:

a $5(x + 3) = 2(3x + 12)$ **b** $10 - 2x = 3x - 5$

c $15 - 2x = 9$ **d** $\frac{6x - 3}{3} = \frac{2x + 4}{2}$

2 Use the following expressions to make as many equations as you can. Solve your equations.
What is the smallest and largest solution you can achieve?

$5x + 7$

$2(4 - 3x)$ $\frac{3x + 1}{2}$

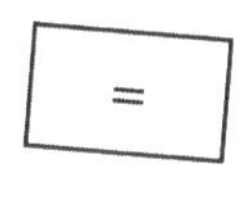

$4 - 8x$ $=$

3 Write an equation to represent each statement. Solve your equation to find the number first thought of:

a I think of a number. Treble the number plus five is equal to double the number plus 10.

b I think of a number. If I multiply it by 5 and subtract it from 18, I get 4.

c I think of a number. Treble the number is twelve more than five times the number.

4 Use an equation to find the missing variable in each case:

a $2x + 12$ $5x + 2$

b

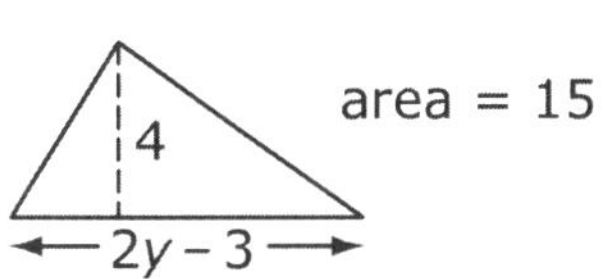

area = 15

c

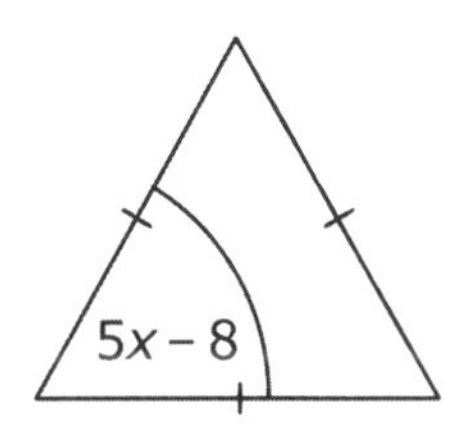

d

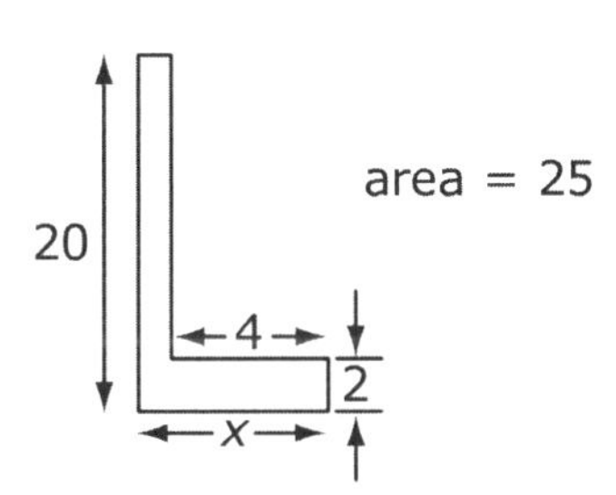

area = 25

5 Solve these equations, giving your answers for x in terms of p, q and r:

a $px + q = r$ **b** $r(x - q) = p$ **c** $px + r = qx - r$

Level 6

a Solve this equation

$7 + 5k = 8k + 1$ *1 mark*

b Solve these equations. Show your working.

i $10y + 23 = 4y + 26$ *2 marks*

ii $\frac{3(2y+4)}{14} = 1$ 2 marks

Level 7

Look at these expressions.

$n - 2$	$2n$	n^2	$\frac{n}{2}$	$\frac{2}{n}$

a Which expression gives the greatest value when n is **between 1 and 2**? *1 mark*

b Which expression gives the greatest value when n is **between 0 and 1**? *1 mark*

c Which expression gives the greatest value when n is **negative**? *1 mark*

S2.1HW Metric units of area

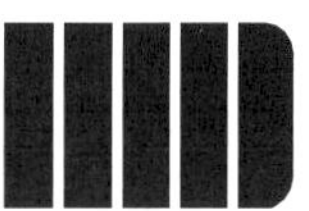

Remember:

- Area of triangle = $\frac{1}{2}$ x base x height
- Area of parallelogram = base x perpendicular height

1 Find the areas of these compound shapes:

a

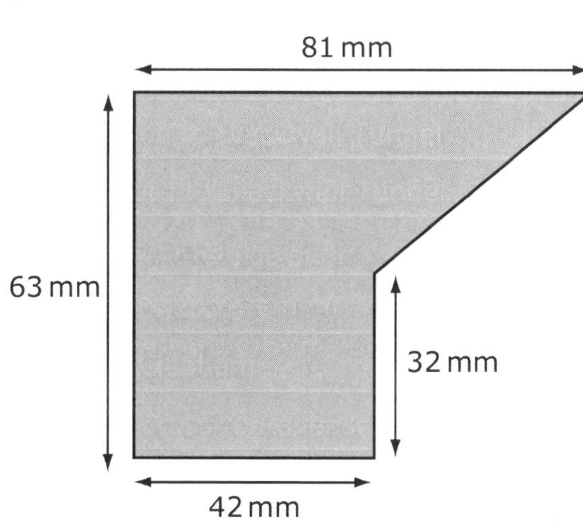

b

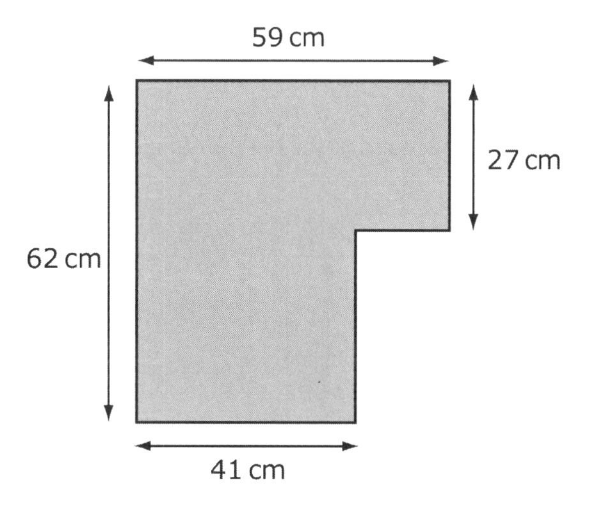

c

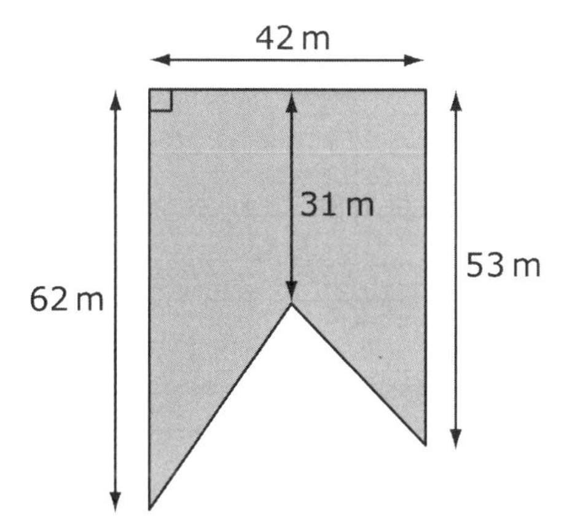

d

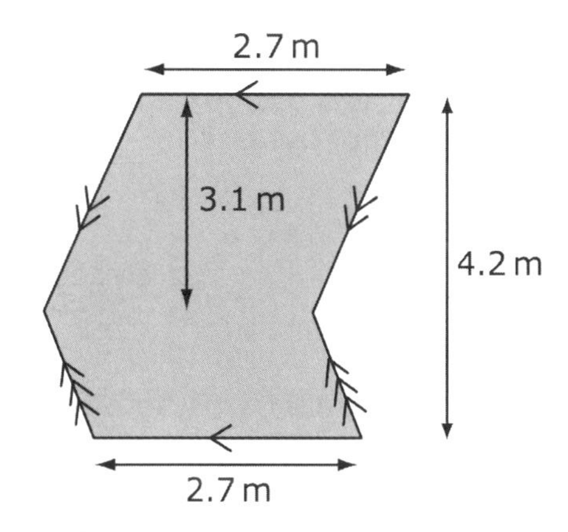

2 Convert your answers to question 1 into these units:

a cm^2

b m^2

c km^2

d mm^2

S2.2HW More metric units

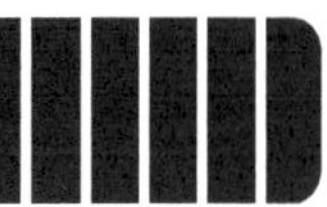

Remember:
- Volume of a cuboid = ℓbh
- $1\text{ ml} = 1\text{ cm}^3$

1 Find the volume of this hut

a in cm^3

b in m^3.

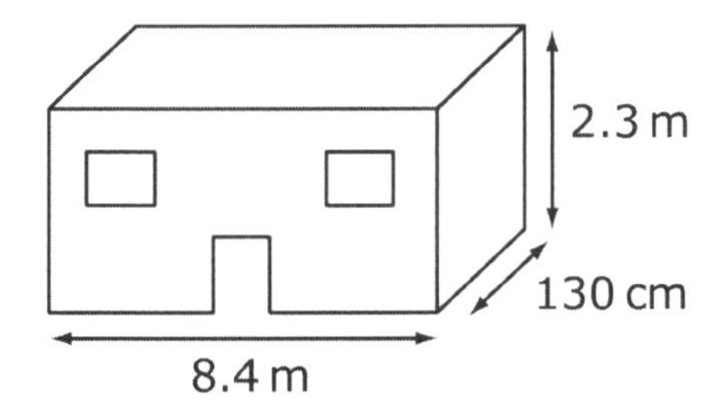

2 Find the volume of these solids:

a

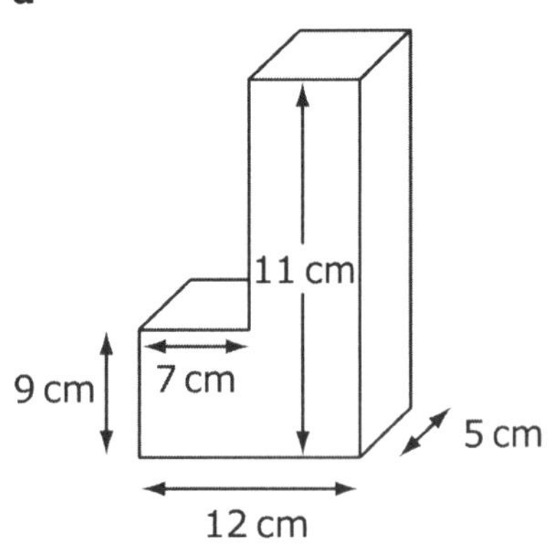

b

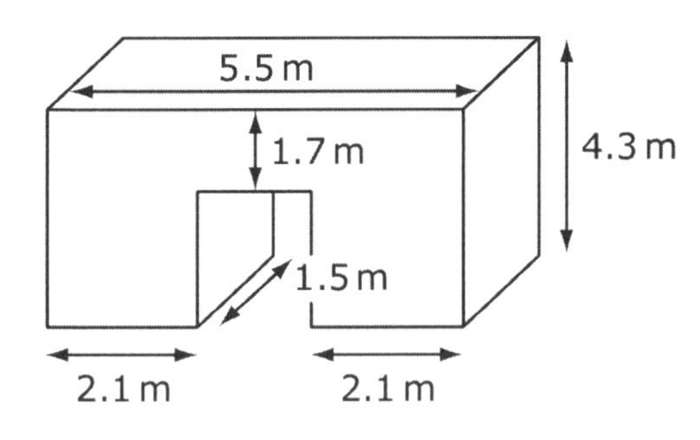

3 The volume of this gift box is 240 cm^3.
Find the length of the box.

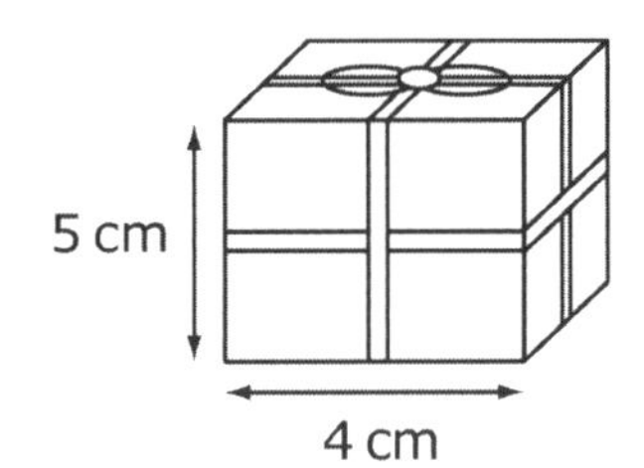

4 Will the contents of this bottle of oil fit into this container?

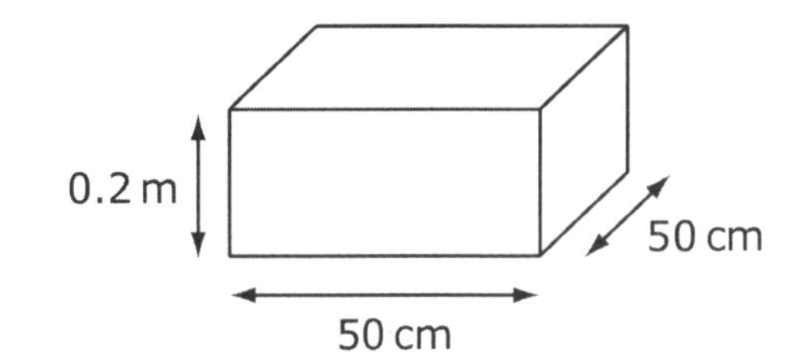

S2.3HW Circumference of a circle

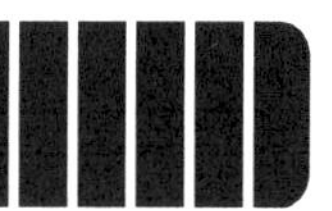

Take $\pi = 3.14$ in this exercise or use the π key on your calculator.

1 A length of wire is bent to form a square of side 5 cm. The same length of wire is then bent into a circle. What is the radius of this circle?

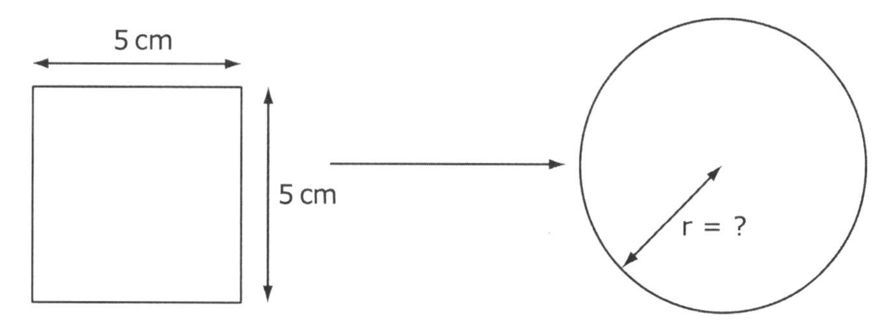

2

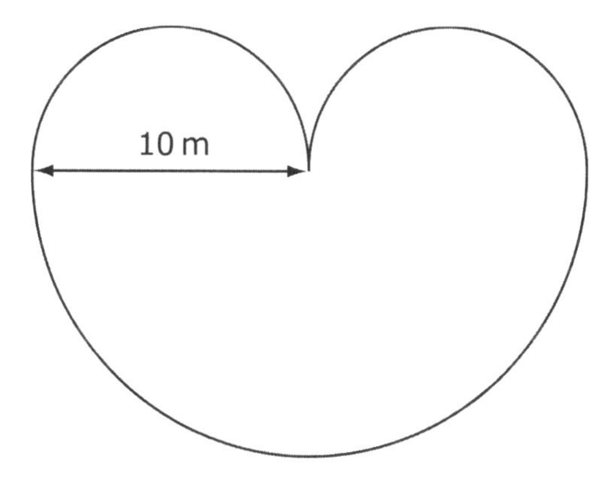

A boating pool is designed in the shape shown above. Find the perimeter of the pool.

3 A playground consists of an L shape with a quarter circle. Find the perimeter of the playground.

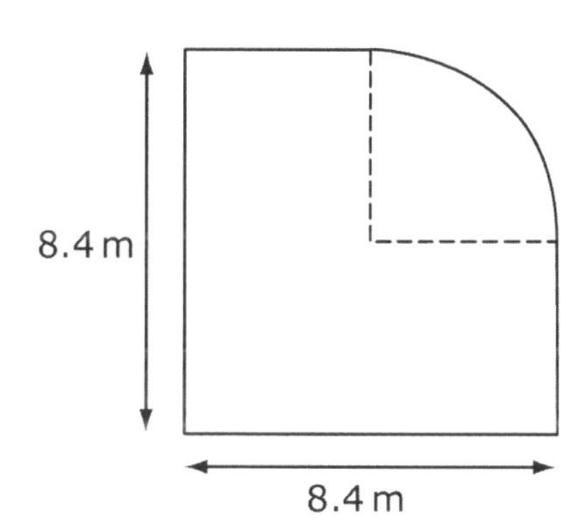

S2.4HW Area of a circle

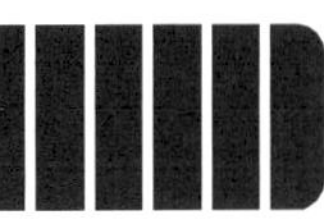

Take $\pi = 3.14$ or use the π key on your calculator.

1 Find the area of each shape.

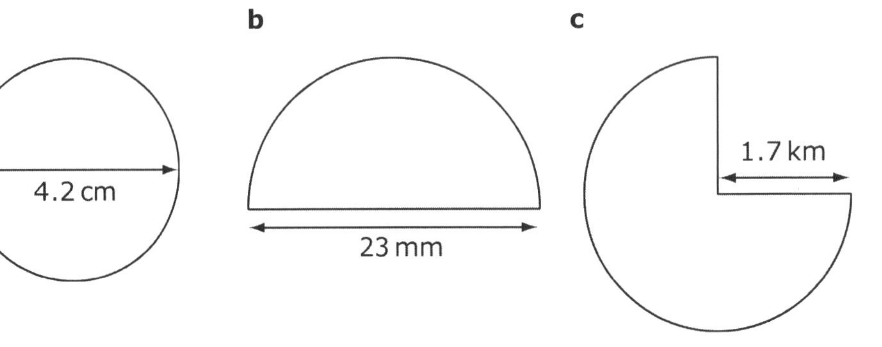

2 Find the radius of a circle with area $24.3\,\text{cm}^2$.
Give your answer to 1 d.p.

3 Find the diameter of a circle with area $42.9\,\text{m}^2$.
Give your answer to 1 d.p.

4 A circular swimming pool has circumference 87.5 m.

a Find the radius of the pool (to 1 d.p.).

b Find the area of the surface of the pool (to 1 d.p.).

5 Find the area of this playing field.

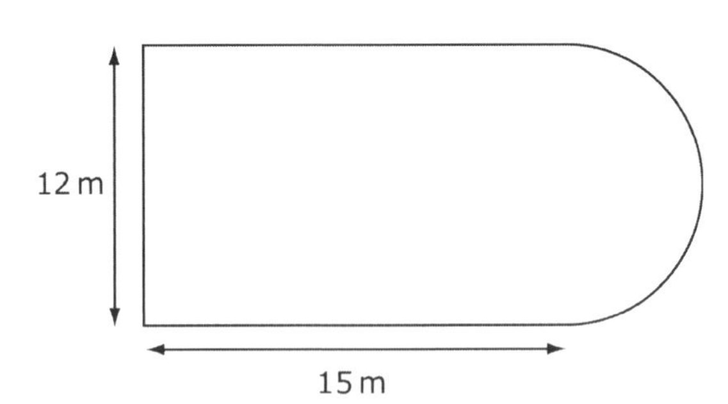

6 A square with sides 5 cm has the same area as a circle. What is the radius of the circle (to 1 d.p.)?

Finding volumes

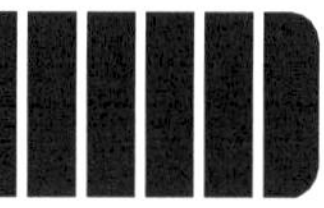

Remember:
Volume of a prism = area of cross-section x length

1 Find the volume of a cube of side 0.4 m in
a cm^3 **b** m^3.

2 Find the volume of these compound shapes.

a

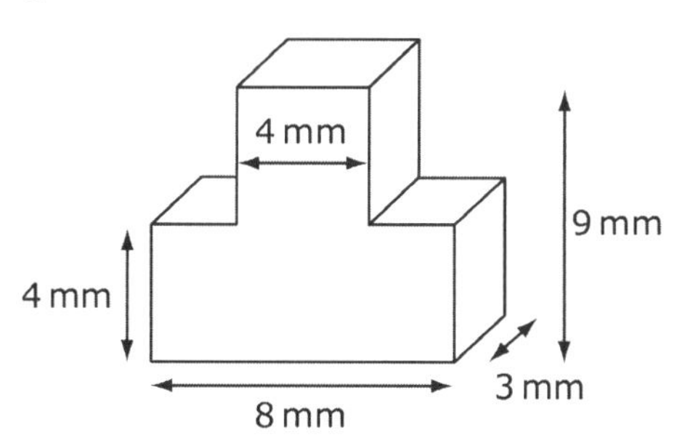

b

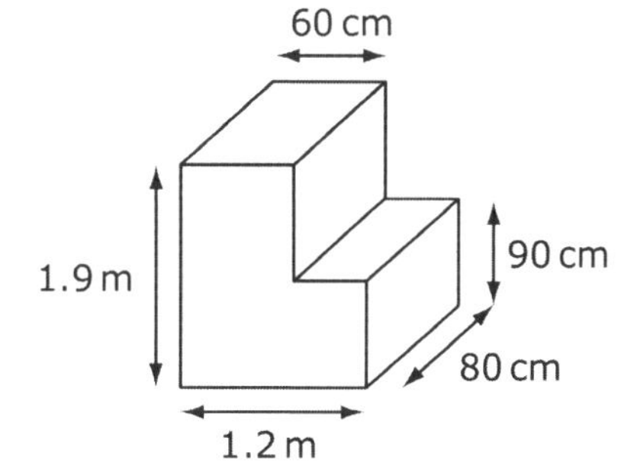

3 Dilip is making a concrete ramp for a wheelchair. What volume of concrete does he need?

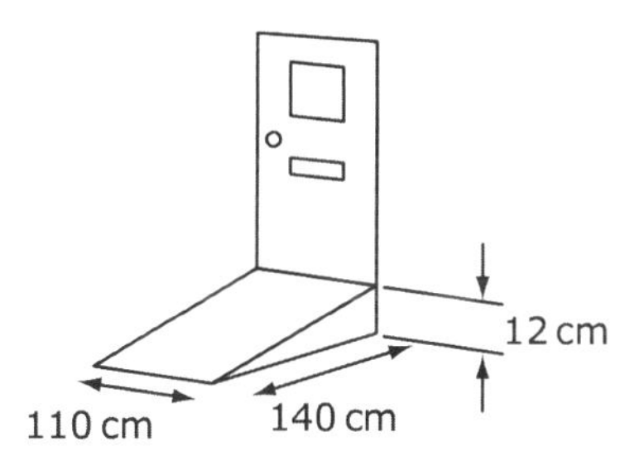

4 Sally has built a coal bunker. What is the volume of the coal bunker?

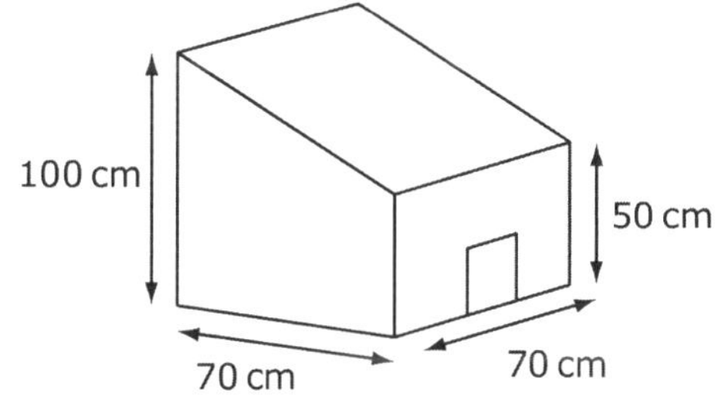

Hint:
Area of trapezium = $\frac{1}{2}(a + b)h$

5 Find the volume of this cake tin.

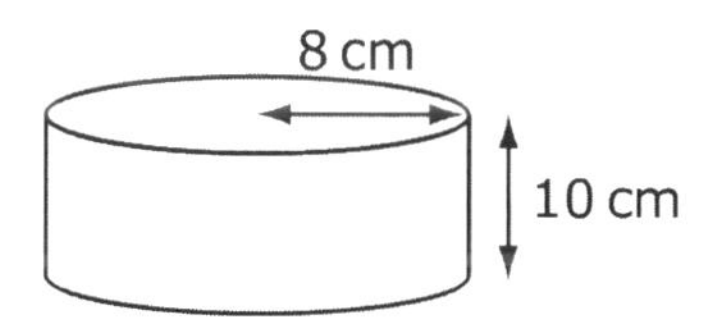

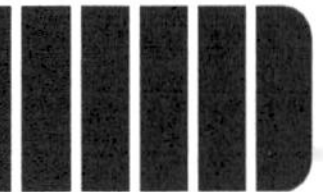

S2.6HW Surface area

1 Find the surface area of a cube of side 7 cm.

2 Find the surface area of this box:

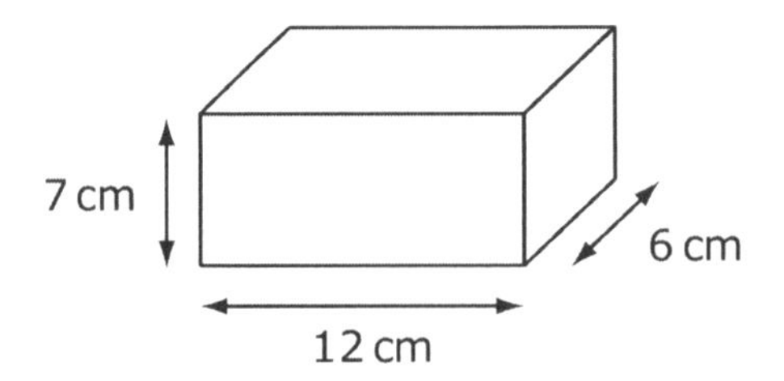

3 Find the surface area of this triangular prism.

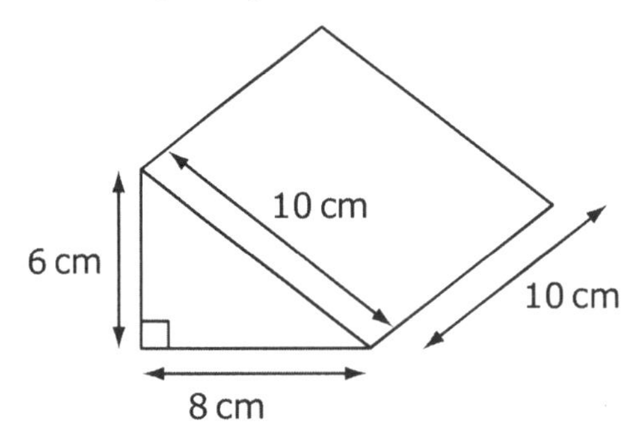

4 Which of these three solids has the greatest surface area?

a

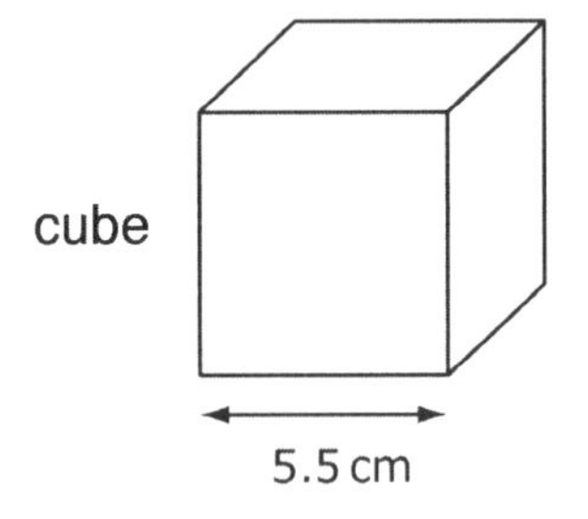

b

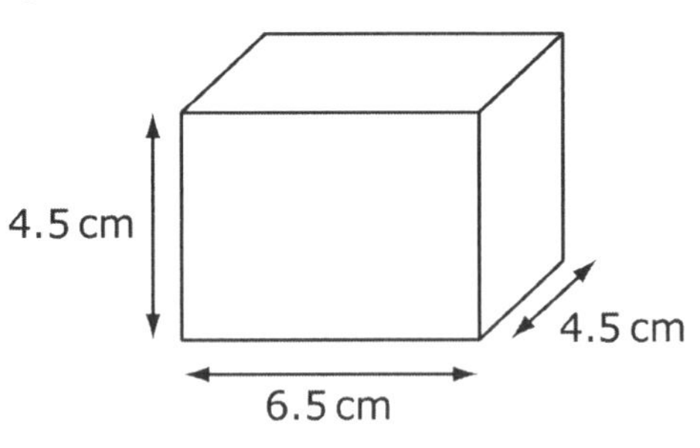

c

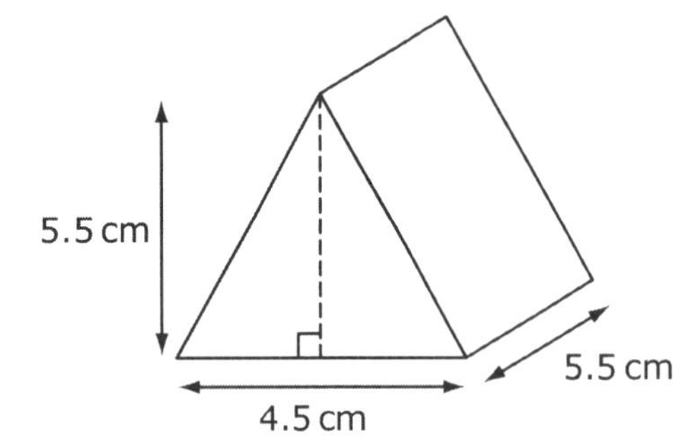

Level 6

A trundle wheel is used to measure distances.

Imran makes a trundle wheel, of **diameter 50 cm**.

a Calculate the **circumference** of Imran's trundle wheel.
Show your working. *2 marks*

b Imran uses his trundle wheel to measure the length of the school car park.
His trundle wheel rotates **87 times**.
What is the **length** of the car park, to the **nearest metre**? *1 mark*

Level 7

At Winchester there is a large table known as the Round Table of King Arthur.

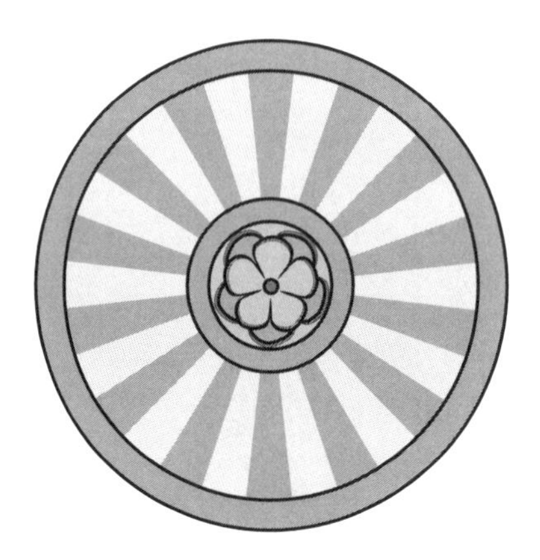

The **diameter** of the table is **5.5 metres**.

a A book claims that 50 people sat around the table.

Assume each person needs 45cm around the circumference of the table.
Is it possible for 50 people to sit around the table?

Show your working to explain your answer. *3 marks*

b Assume people sitting around the table could reach only **0.5 m**.

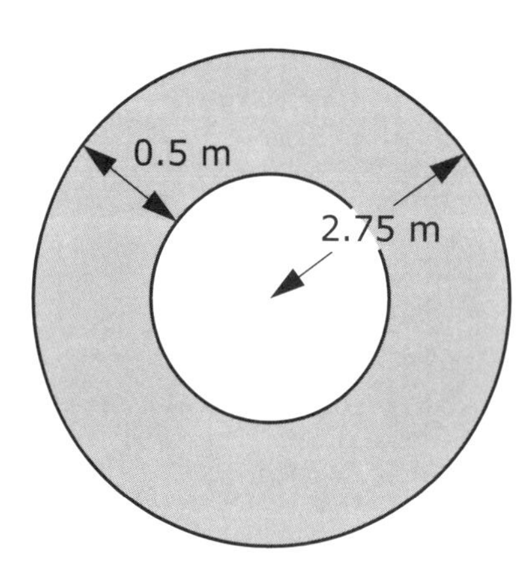

Calculate the **area** of the table that could be reached.

Show your working. *3 marks*

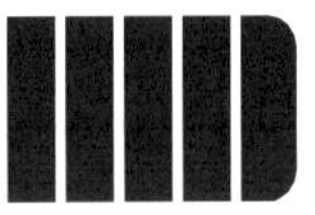

A3.1HW Straight-line graphs

1. Match each graph with its equation:

a **b** **c** **d** **e**

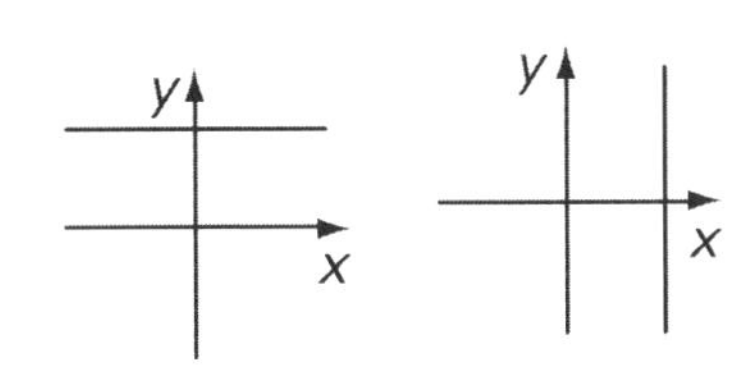

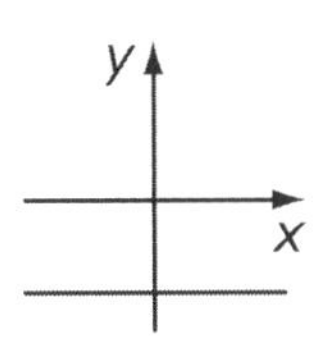

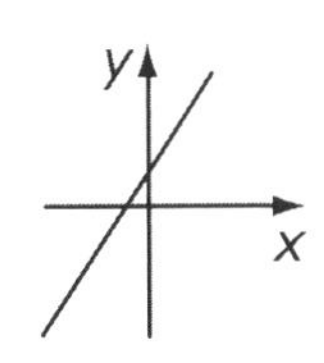

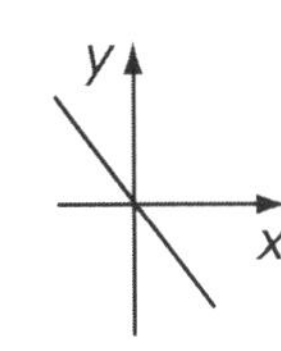

$x = 2$	$y = 2x + 1$	$y = 6$	$y = -3x$	$y = -3$

2. For each equation, complete a table of coordinates. Plot the graph on a set of axes from −10 to +10.

For example, $y = 3x - 2$

x	1	2	3
y	1	4	7

a $y = 2x + 2$ **b** $y = 4x - 3$ **c** $y = 5$ **d** $y = 8 - 2x$

3. **a** Where do the graphs $y = 3$ and $x = 5$ intersect?

b Write the equations for two graphs that intersect at $(4, {}^{-}2)$.

4. Find the equations of the graphs with these tables of coordinates:

> **Hint:** It may be helpful to use differences, when finding the general term of a sequence.

a

x	1	2	3	4
y	1	3	5	7

b

x	3	3	3	3
y	17	4	${}^{-}2$	5

5. Select equations(s) from the box to fit the following descriptions:

a A vertical line.

b A graph that is not straight.

c A pair of parallel lines.

d A line with a negative direction.

e A pair of perpendicular lines.

$y = 7$	$y = -2x + 1$
$y = 3x + 1$	$y = x^2$
$x = {}^{-}2$	$y = 2 + 3x$

A3.2HW Gradient of diagonal lines

1 Match the gradient with the line.

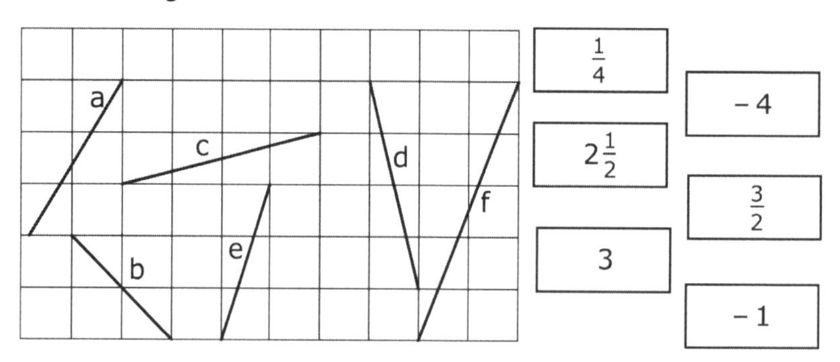

2 Use the gradient formula to find the gradient of these line segments.

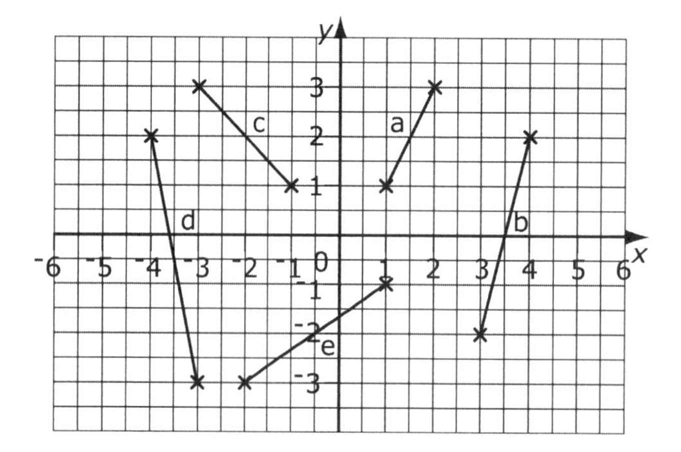

Hint:
m = gradient.

$$m = \frac{y_2 - y_1}{x_2 - x_1}$$

3 **a** This is a square ABCD. Find the gradients: AB, BC, CD and DA.

b What do you notice about gradients of opposite sides? Why is this?

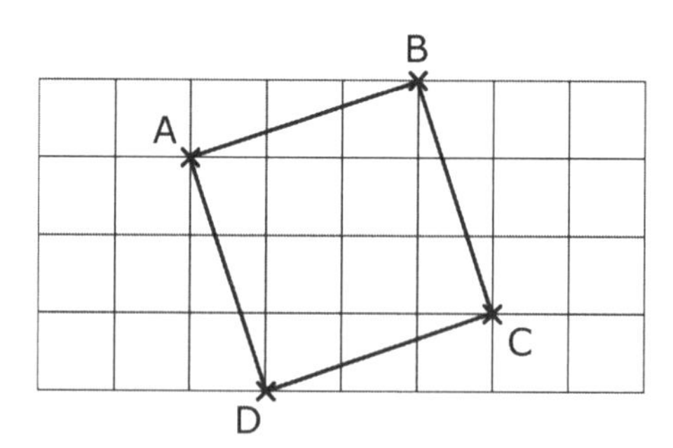

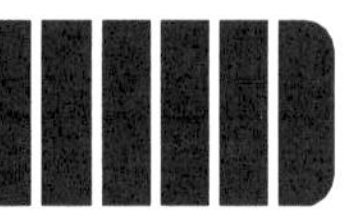

A3.3HW The equation of a straight line

1 Match each graph with its equation.

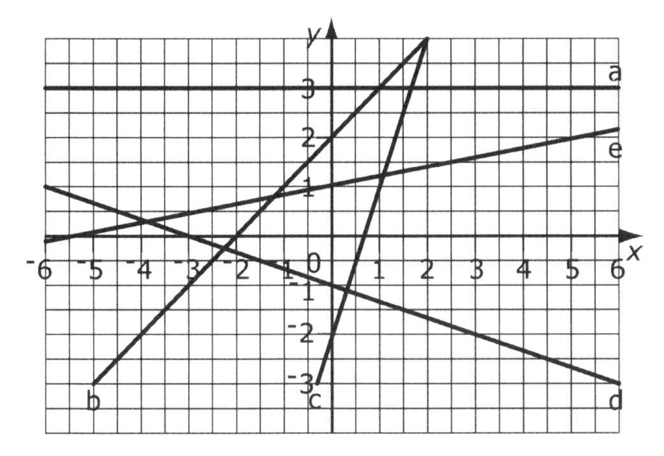

$y = 3$

$y = -\frac{1}{3}x - 1$

$y = \frac{1}{5}x + 1$

$y = x + 2$

$y = 3x - 2$

2 The line graph shown has equation $y = 3x + c$

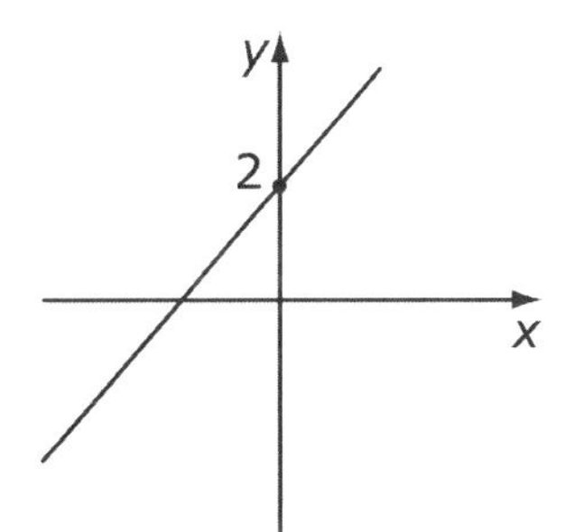

a What is the value of c?

b Will the graph pass through the point (3, 10)?

c Write the equation of a graph parallel to the one shown.

3 Find the equation of the line described in each case:

a Gradient 3, cuts through (0, 6).

b Steepness of 4, passes through (0, $^{-}2$).

c Parallel to $y = 5x - 1$, cuts the y-axis at (0,10).

d Cuts through (0, 7) and (5,17).

4 Here are three equations.

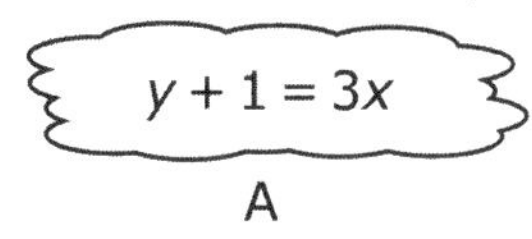

A

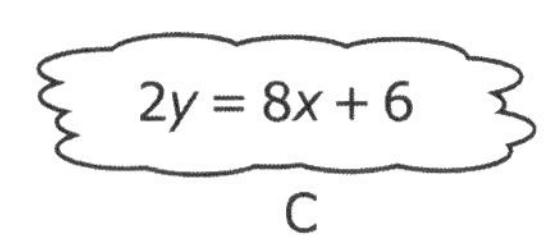

B

$2y = 8x + 6$

C

Find the odd one out in each case:

a Gradient is $^{\pm}3$

b Slopes in a positive direction

c Cuts the y-axis at (0, 3).

5 A line passes through the point (10, 10). Its equation is $y = mx + 4$. Find the value of m.

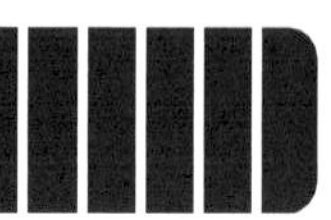

A3.4HW Parallel and perpendicular lines

1 From each box pair up gradients of lines that are perpendicular to each other.

Box A

$\frac{1}{3}$ $\quad ^{-}0.75$ $\quad ^{-}0.\dot{3}$ $\quad \frac{^{-}1}{4}$ $\quad 2\frac{1}{2}$ $\quad \frac{3}{4}$ $\quad ^{-}0.1$ $\quad \frac{^{-}2}{3}$

Box B

4 $\quad ^{-}10$ $\quad \frac{3}{2}$ $\quad \frac{^{-}4}{3}$ $\quad \frac{^{-}2}{5}$ $\quad 3$ $\quad ^{-}3$ $\quad 1\frac{1}{3}$

2 True or false:

a $y = 4x$ and $y = {}^{-}4x$ are perpendicular.

b $y + 3x = 0$ and $3y = x$ are perpendicular.

3 A trapezium, rectangle, kite and parallelogram are to be drawn on to a squared grid in different orientations.

Copy the diagrams.

From the box, choose suitable gradients for the edges shown. You must use each value once and only once.

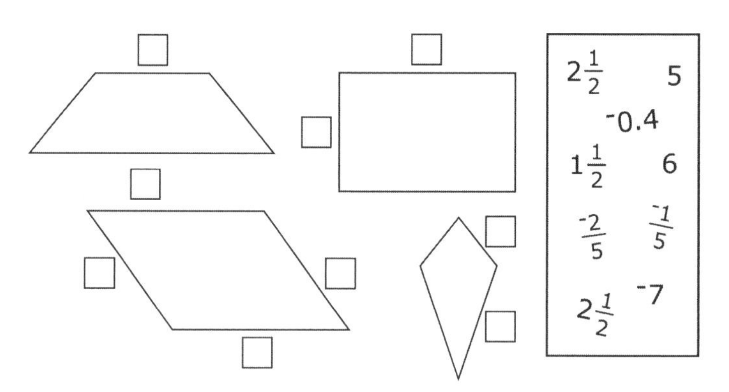

4 Explain why $y = \frac{1}{2}x$ and $y = {}^{-}3x$ cannot be equations of the diagonals of a kite.

5 Find the equations for lines a and b.

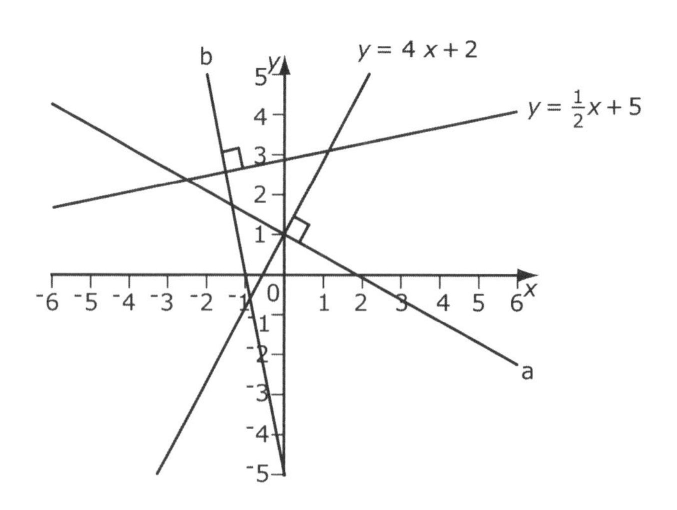

A3.5HW Curved graphs

1 For each equation given, copy and complete this table, and then use it to plot a graph of the equation on appropriate axes.

x	$^{-}4$	$^{-}3$	$^{-}2$	$^{-}1$	0	1	2	3	4
y									

a $y = x^2 - 1$ **b** $y = 2x^2$ **c** $y = x^3$

2 Ben and Clare each drew a graph.

Explain how you know they have made a mistake in their work.

Ben

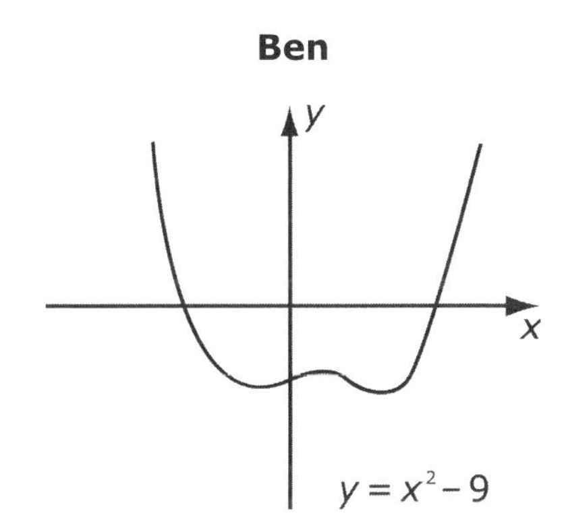

Clare

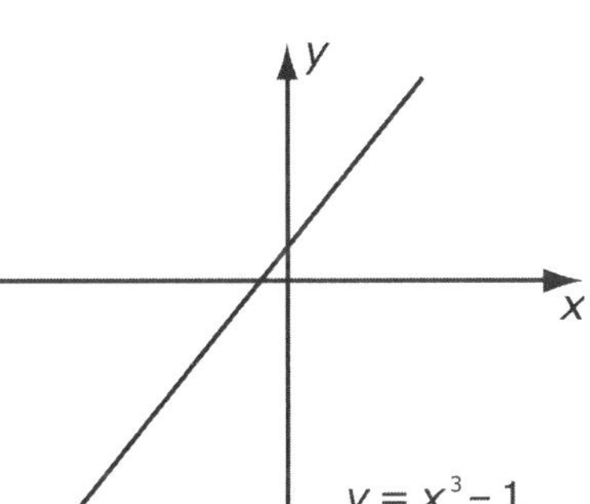

3 Match these graphs and equations.
Sketch the graph of the remaining equation.

a

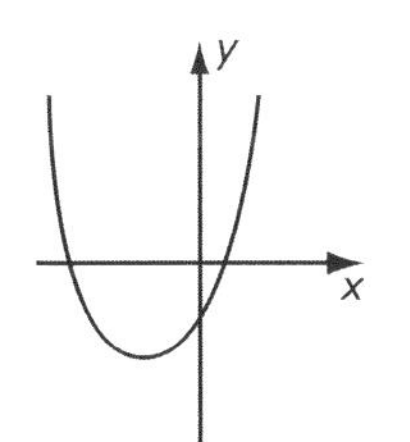

b

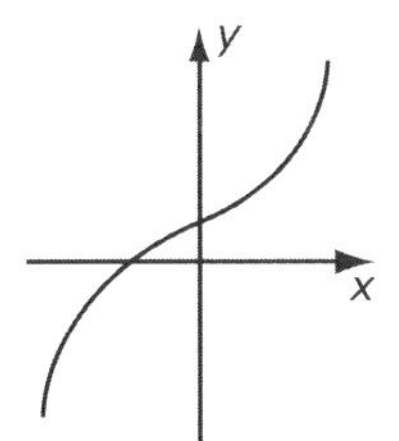

c

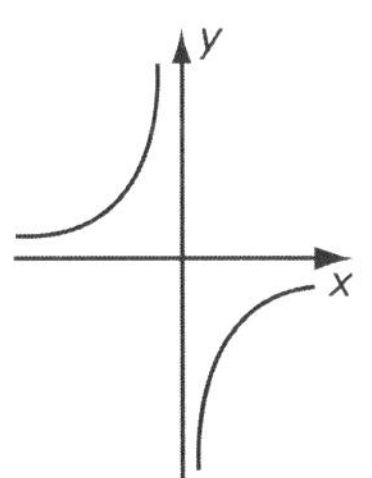

d

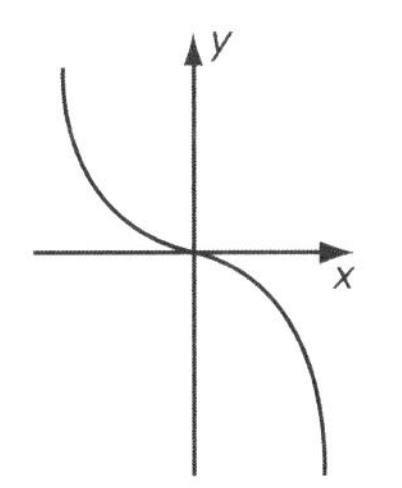

e

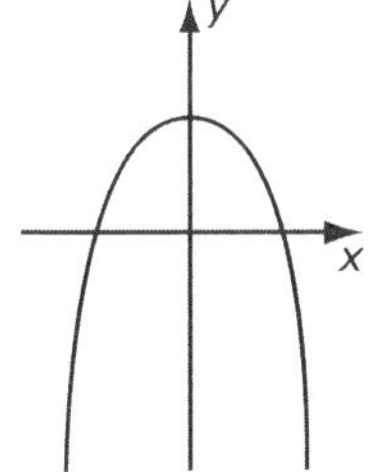

$y = x^3 + 4$	$y = -\frac{1}{x}$	$y = {}^{-}x^3$	$y = x^2 - 2x - 3$	$y = 5 - x^2$	$y = 3x + 4$

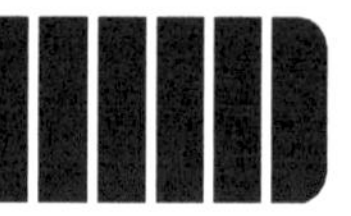

A3.6HW Interpreting graphs

1 The graph shows a distance–time graph for two spiders on a wall.

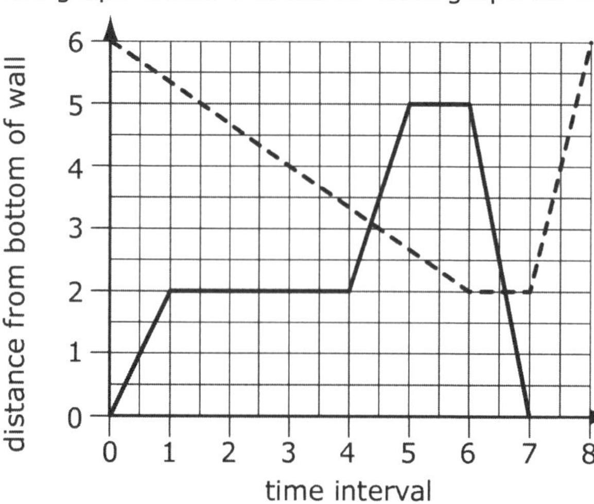

a For how long did the male spider stop during his journey?

b Which spider travelled furthest during their journey?

c True or false: in the first section of the journey, the male spider moved at a speed of 2 cm/s. If false give the correct speed in cm/s.

d What happened after $4\frac{1}{2}$ minutes?

e After 5 minutes, which spider was closest to the floor?

2 **a** Construct a distance–time graph for this journey, on a copy of these axes.

A car leaves Sutton Coldfield at 12 noon and travels 10 miles to the M6 motorway, keeping to the 30 mph speed limit.
A traffic jam halts its journey fo 20 minutes but it then manages to travel at 40 mph for $\frac{1}{2}$ hour. At this point, the motorway clears and the driver sticks to 70 mph, driving steadily until 2 o'clock.

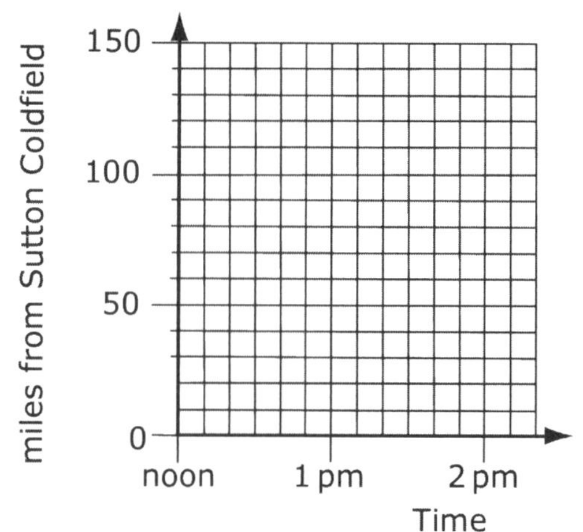

b How far is the car from Sutton Coldfield at 2pm?

c Calculate the car's average speed over its whole journey.

Level 6

I went for a walk.

The distance-time graph shows information about my walk.

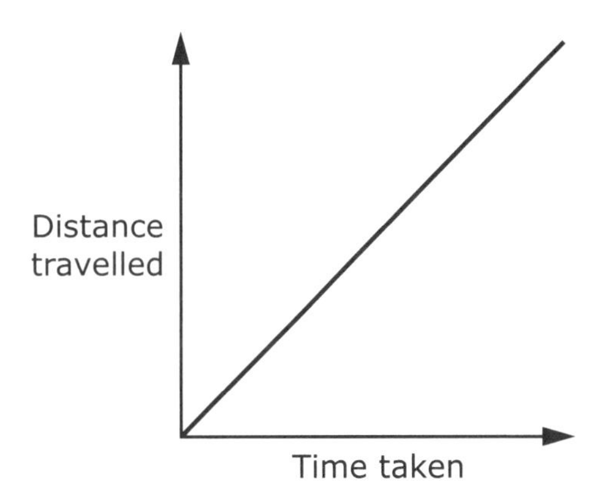

Write down the statement below that describes my walk.

- I was walking faster and faster.
- I was walking slower and slower.
- I was walking north-east.
- I was walking at a steady speed.
- I was walking uphill.

1 mark

Level 7

Here are six different equations, labelled A to F

A	$y = 3x - 4$	**B**	$y = 4$	**C**	$x = -5$
D	$x + y = 10$	**E**	$y = 2x + 1$	**F**	$y = x^2$

Think about the graphs of these equations.

a Which graph goes through the point **(0, 0)**? *1 mark*

b Which graph is **parallel** to the y-axis? *1 mark*

c Which graph is **not** a **straight line**? *1 mark*

d Which **two** graphs pass through the point **(3 , 7)**? 2 marks

e The diagram shows the graph of the equation $y = 4 - x^2$.

What are the coordinates of the points where the graph of this equation meets the graph of equation **E?**

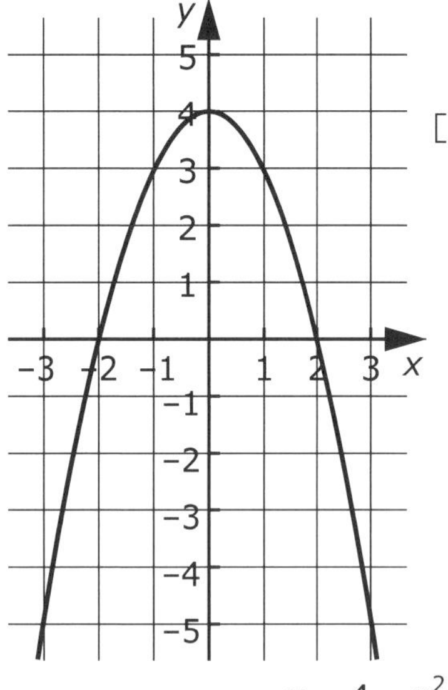

$y = 4 - x^2$

3 marks

N3.1HW Multiplying by powers of 10

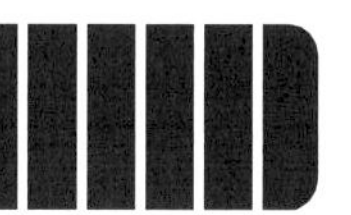

1 Write each of the following as ordinary numbers:

a 10^2 **b** 10^6 **c** 10^{-1} **d** 10^{-3}

2 Calculate, using a mental method:

a 3×0.1

b 0.01×6

c 70×0.001

d $8 \div 0.1$

e $90 \div 0.1$

f $7 \div 0.001$

3 Work out which 'power of 10' is missing in each calculation below.

a $3.1 \times \square = 0.31$

b $7 \div \square = 700$

c $\square \times 0.71 = 710$

d $80 \div \square = 0.8$

e $6.21 \times \square = 621\,000$

f $7.9 \div \square = 0.0079$

N3.2HW Rounding

1 Copy and complete the table below, writing the numbers given to the degree of accuracy shown at the top of each column.

Number	1 d.p.	2 d.p.	3 d.p.	4 d.p.
2.3432				
0.54678				
12.9531				
5.9256				
1.3862				
54.6257				
5.00391				
0.000 056 56				
0.000 052				

2 Work out an approximate answer to these calculations.

a 3.912 x 3.015

b 15.85 ÷ 4.02

c 23.93 x 6.732

d 14.93 ÷ 5.97

3 Steven says that 3 miles is roughly equivalent to 5 kilometres. Maya says that 1 mile equals 1.609 kilometres.

a Using both conversions, find out how many kilometres is equivalent to a journey of 70 miles. Give your answers to 3 decimal places.

b For 1 mile, find the **difference** between the number of kilometres using both Steven's and Maya's conversions. Give your answer to 2 decimal places.

N3.3HW Adding and subtracting

1 Work out these using a mental method.

a $^{-}8.6 - {}^{-}3.4$

b $2.9 - 4.9 - {}^{-}1.2$

c $1.7 + 2.6 - 1.3$

d $^{-}3.1 + {}^{-}2.9$

e $^{-}0.48 + {}^{-}0.003$

f $0.6 - {}^{-}0.21 - 0.3$

> **Remember:**
> The standard column method:
>
> $$\begin{array}{rr} & 0.41 \\ & 3.6 \\ & 15 \\ + & 3.27 \\ \hline & 22.28 \\ \hline \end{array}$$
>
> Make sure you write the numbers with their decimal points lined up.

2 Use a standard column method to calculate:

a The total length of 6 pieces of ribbon of lengths:
2.03 m, 0.4 m, 93 cm, 4.1 m, 0.563 m, 84 cm.

b Two pieces of string of lengths 3.1 m and 41 cm joined end to end and then 1.431 m cut off this new length.

3 Write down the calculations you would do to estimate the answers to question 2.

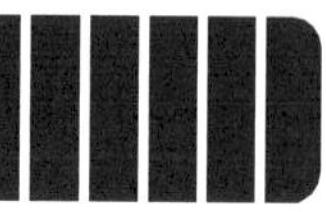

Puzzle

In each of the diagrams:

- The number in the triangle is the **product** of the numbers in the two circles.
- The number in the square is the **sum** of the numbers in the two circles.

Copy and complete each diagram.

Use a mental method for your calculations.

1

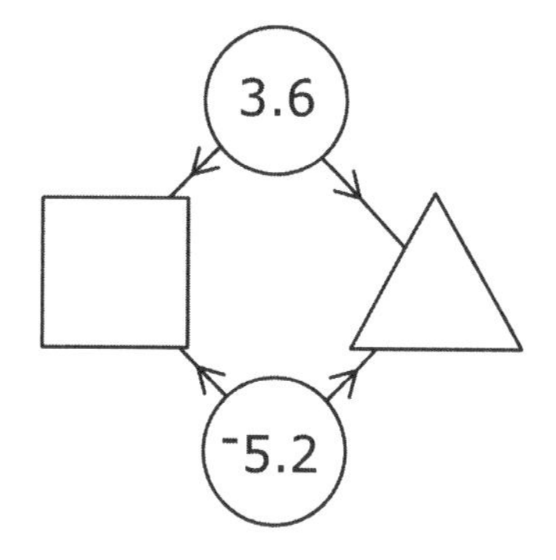

2

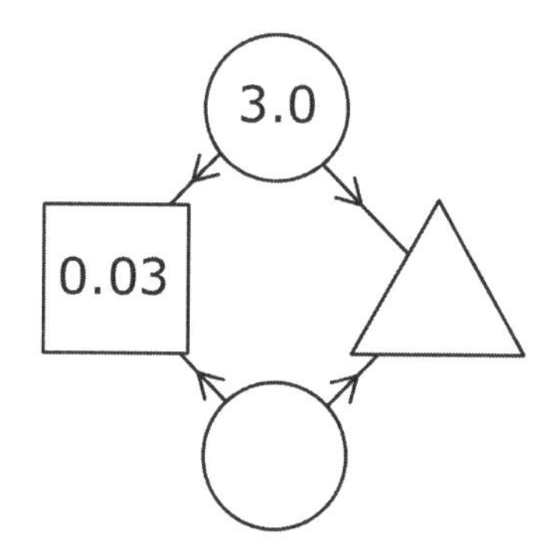

3

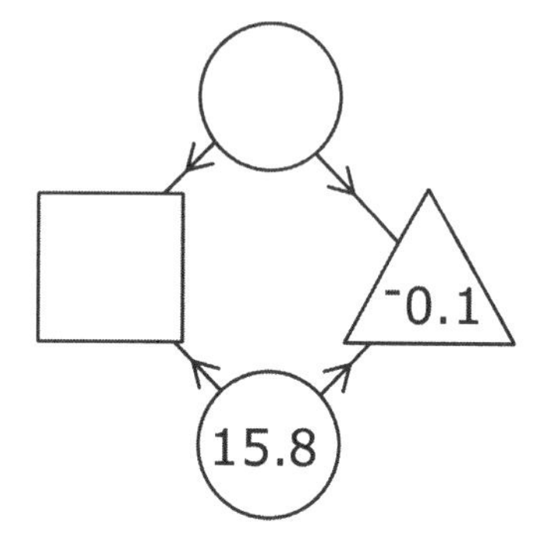

4

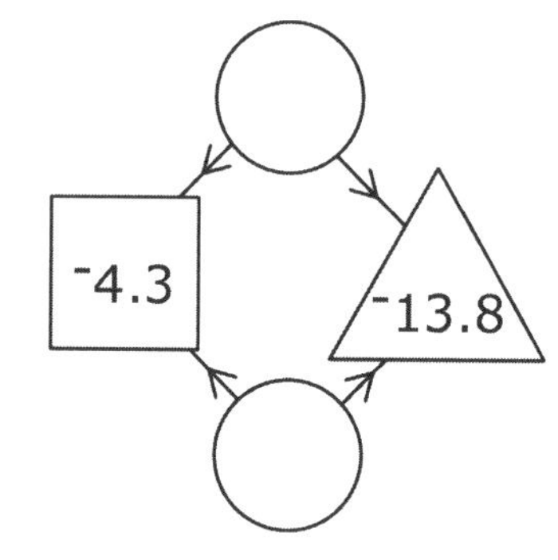

N3.5HW Multiplying decimals

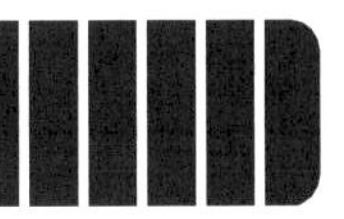

Alison lives in Macclesfield and works in Manchester. Her journey to work each day is 28 miles. She is trying to decide whether it is better to travel to work by car or by train. Here are some ideas she has written down.

Travelling by car …

Car insurance = £524.32 (each year)
Road Tax = £180 (yearly)
Servicing = £240 (twice a year)

Fuel costs

Petrol = £71.9p per litre
Consumption = 41 miles per gallon
(1 gallon ≈ 4.75 litres)

Travelling time

Each journey lasts about 55 min

Travelling by train …

Monthly season ticket = £165.42
Daily return ticket = £10.25

Travelling time

Home to station = 15 min walk
Journey on train = 36 min
Station to work = 8 min walk

Alison works for 44 weeks a year. She has a four-week holiday in August.
Write a short report recommending which form of transport Alison should take. Explain and justify your answer.

N3.6HW Dividing decimals

1 For each of these divisions:

- Estimate an approximate answer.
- Work out the exact answer using either a mental or written method.

Give your answer to **d**, **e** and **f** correct to 2 d.p.

a $1582 \div 0.4$

b $1.134 \div 0.09$

c $13.14 \div 0.15$

d $1.632 \div 0.7$

e $0.137 \div 0.4$

f $0.00483 \div 0.013$

2 Henrietta's father won £29 342 on the National Lottery.
Henrietta said that if he shared it equally between the 42 people in her street they would each get £6986.19.

a Without working out the exact answer, how do you know that she was wrong?

b What error do you think she may have made?

c What is the exact amount each would receive (to the nearest penny)?

3 Sam is running in a marathon.
The total distance of the course is 25.8 km.
Sam runs at an average of 8.5 km per hour.

How long will it take Sam to complete the marathon?
Give your answer in hours as a decimal to 2 d.p.

N3.7HW Ratio

1 Investigation

£1500 is to be split between three sisters, Amanda, Jane and Louise.

Amanda must have less than $\frac{1}{3}$ of the total.

Jane must have more than $\frac{1}{3}$ of Louise's share.

Louise must have less than 160% of Amanda's share.

a Find a way of splitting the money using a whole number ratio so that it satisfies all of these conditions.

b Investigate other possible answers.

2 a The ratio of the areas of these two rectangles is 2 : 9. Find the volumes of a and b.

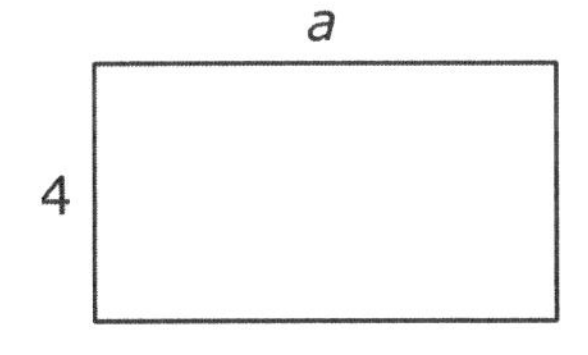

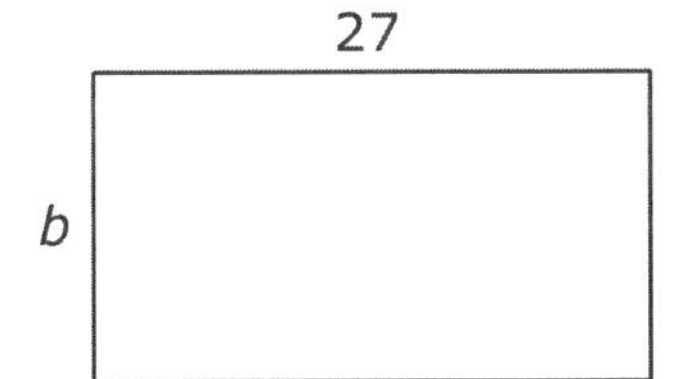

b The ratio of the areas of these two triangles is 3 : 5. Find the values of x and y.

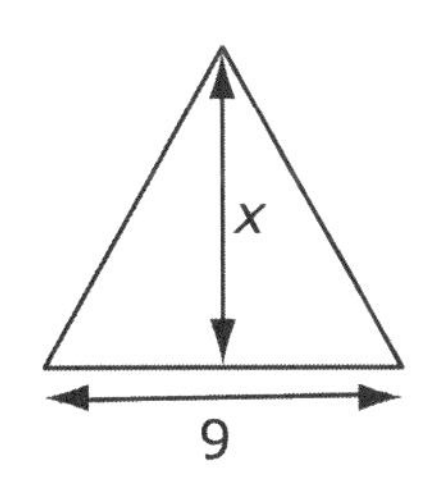

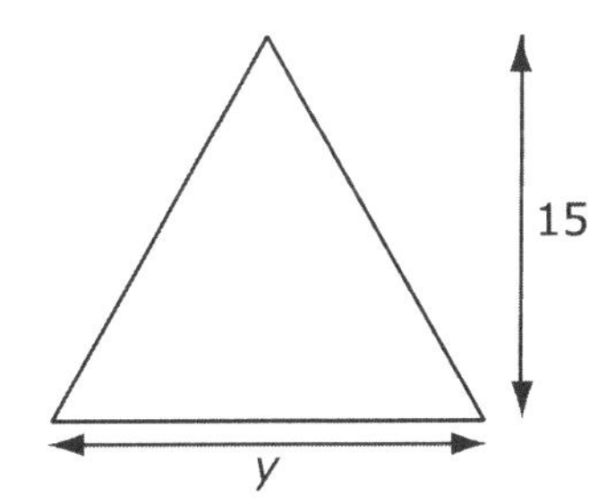

N3.8HW Multiplicative relationships

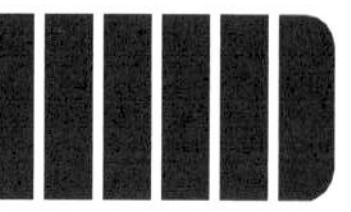

1 The table shows the increase in price of six items between 1993 and 2003.

Item	**Price in 1993**	**Price in 2003**	**Price increase** (scale factor)
CD	£12	£15	× 1.25
Car	£10 000		× 1.5
Cricket bat	£20	£35	
Computer	£1000	£650	
Mobile phone	£80		× 0.75
Paperback novel	£5		× 1.6

a Copy and complete the table.

b Sarah says: 'CDs now cost 125% of their cost in 1993'.
Sally says: 'In 1993, CDs used to cost 80% of their current price'.
Who is correct? Explain and justify your answer.

2 A superstore sets its prices in its York and London stores in the ratio 1 : 1.3 respectively.

a Copy and complete the table giving prices to the nearest penny where appropriate.

York store price	**London store price**
£30.56	
	£45
£98.46	
	£212
£70.12	

b Explain how you worked out the York prices.

N3.9HW Direct proportion

1 The table shows how much Dilip earned for working shifts of different lengths.

Hours worked	2.5	6	7	9	11	14
Amount earned	£15.50	£37.20	£43.40	£55.80	£68.20	£86.80

a Are the hours worked and amount earned directly proportional? Show how you decided.

b Work out the scale factor that links the hours and the amount earned.

c Write an equation connecting hours worked and amount earned.

2 Paul earns $1\frac{1}{2}$ times as much as Dilip per hour.
How much does Paul earn for working five 7-hour shifts in one week?

3 Dilip gets a 20% pay rise.
How much does he earn for working a 12-hour shift?

Level 6

A report on the number of police officers in 1995 said:

"There were **119 000** police officers. **Almost 15%** of them were **women**."

a The **percentage** was **rounded** to the nearest whole number, 15.
What is the **smallest** value the percentage could have been, to one decimal place?
Write down the correct answer from the ones in the bubble.

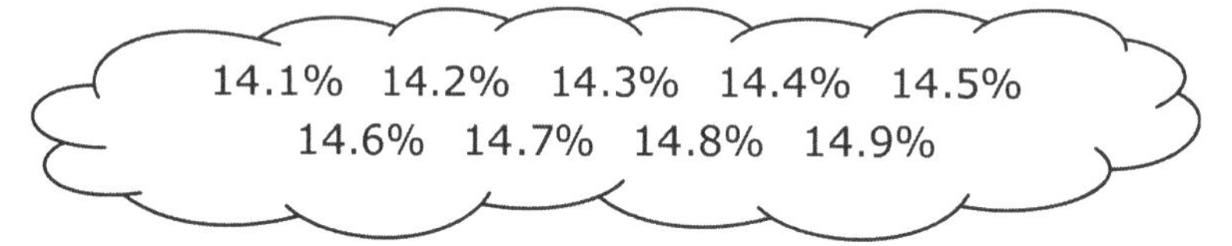

1 mark

b What is the **smallest number** of women police officers that there might have been in 1995?
(Use your answer to part **a** to help you calculate this answer.)
Show your working. *2 marks*

c A different report gives exact figures:

Number of women police officers	
1988	12 540
1995	17 468

Calculate the **percentage increase** in the number of women police officers from 1988 to 1995. Show your working. *2 marks*

d The table below shows the **percentage** of police officers in 1995 and 1996 who were women.

1995	14.7%
1996	14.6%

Use the information in the table to decide which one of the statements below is true. Write down the true statement.

- In 1996 there were **more** women police officers than in 1995.
- In 1996 there were **fewer** women police officers than in 1995.
- There is **not enough information** to tell whether there were more or fewer women police officers.

Explain your answer. *1 mark*

Level 7

The table shows the average weekly earnings for men and women in 1956 and 1998.

	1956	**1998**
Men	£11.89	£420.30
Women	£6.16	£303.70

a For **1956**, calculate the average weekly earnings for women as a percentage of the average weekly earnings for men.

Show your working and give your answer to 1 decimal place.

2 marks

b For **1998**, show that the average weekly earnings for women were a **greater proportion** of the average weekly earnings for men than they were in 1956. *2 marks*

Congruence and transformations

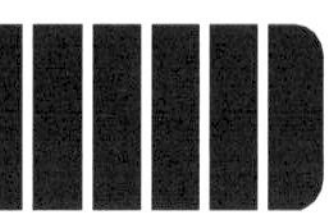

1 The diagram shows transformations of a shape.

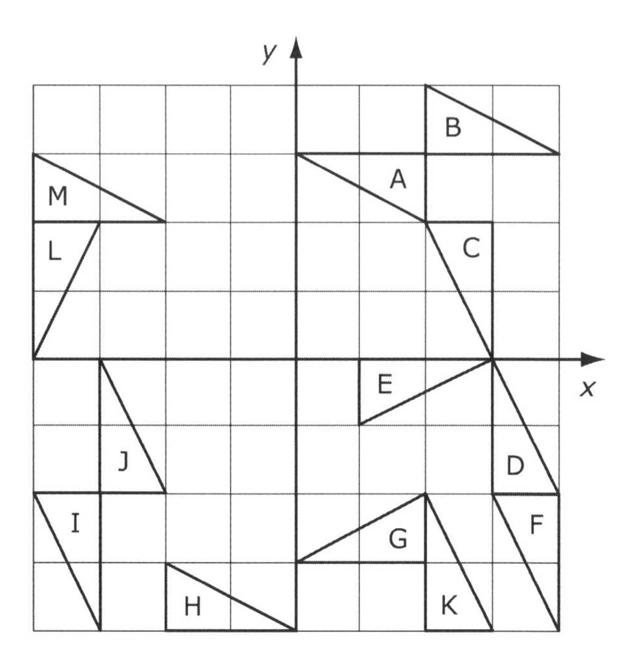

Describe fully the **single** transformation that maps:

a C to F **b** L to M **c** H to M

d A to G **e** I to J **f** D to K

g A to C **h** J to A **i** H to B

j A to H **k** K to M **l** D to F.

2 Draw a grid from $^{-}4$ to $^{+}4$ on both axes.

a Plot points (0, 1), (1, $^{-}1$), (0, $^{-}1$) and join them up. Label the shape A.

b Reflect A in the *y*-axis. Label the new shape B.

c Rotate A 180° about (0, $^{-}1$). Label the new shape C.

d Translate A by $\begin{pmatrix} ^{-}2 \\ 3 \end{pmatrix}$. Label the new shape D.

e Enlarge A by scale factor 2, centre of enlargement (0, $^{-}1$). Label the new shape E.

f Which of B, C, D and E are congruent to A?

g What mathematical word describes A and E?

A and E are s____________ .

S3.2HW Repeated transformations

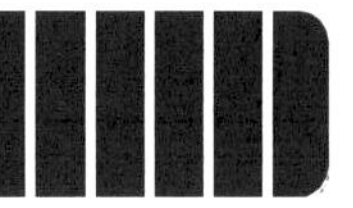

1 Follow the instructions for each diagram below. Make all measurements accurate.

a Copy each diagram. Reflect the flag A in mirror line M_1. Label this A_1.

b Reflect the flag A_1 in mirror line M_2. Label this A_2.

i

ii

iii

c For each diagram, find a relationship between the lengths M_1M_2 and AA_2.

> **Hint:**
> Let the distance $AM_1 = x$ and $M_1M_2 = y$.

2 Copy shape A in the diagram. Follow these instructions to make a tessellation.

i Rotate A 180° about the marked point (•) to get B.

ii Rotate B about a point on the middle of another side of B to get C.

iii Continue the pattern.

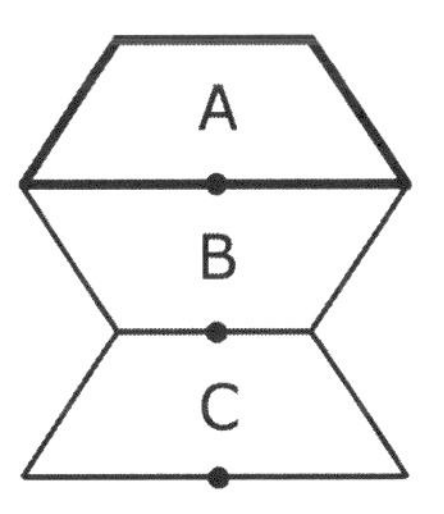

3 Copy each shape. Rotate each shape about the marked point (•) 180° clockwise. (Points of rotation are the midpoint of the side.). Shape **a** is done for you.

a

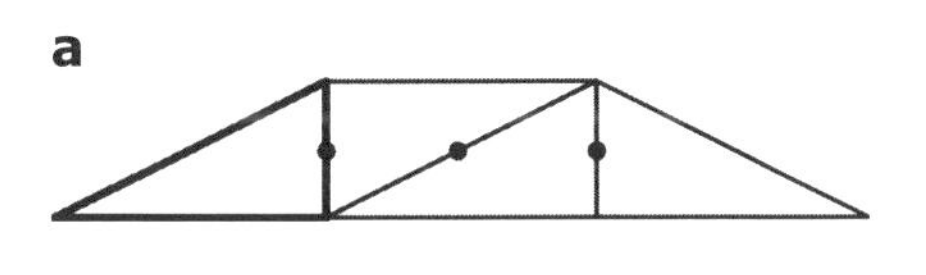

b

c

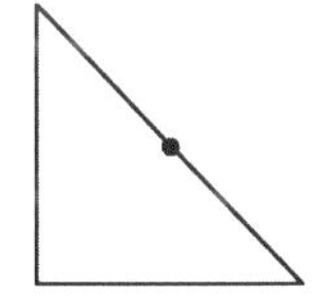

d

e

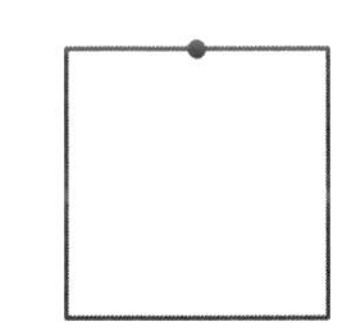

f

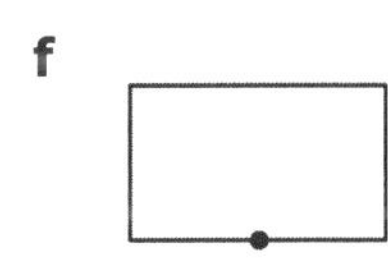

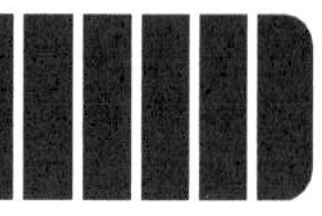

S3.3HW Combining transformations

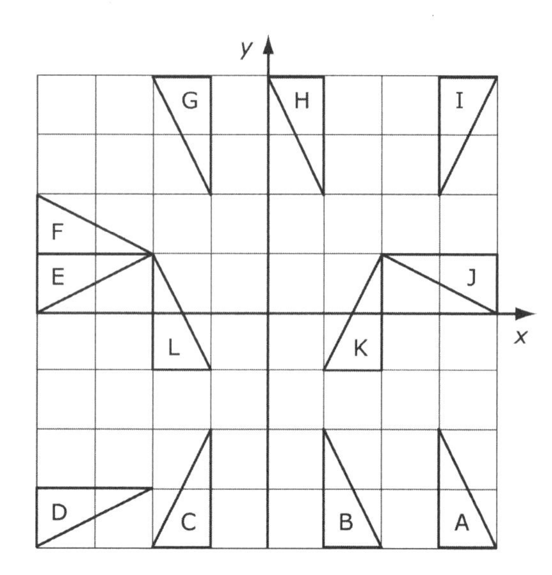

1 **a** Describe fully the single transformation that maps

i B to C **ii** C to A.

b The combination of these transformations can be mapped in one move. Describe fully the single transformation that maps B to A.

2 **a** Describe fully the single transformation that maps

i E to F **ii** F to G.

b The combination of these transformations can be mapped in one move. Describe fully the single transformation that maps E to G.

3 Look at the questions above. Reflecting the shapes in two parallel lines is equivalent to what single transformation?

4 **a** Describe fully the single transformation that maps

i B to C **ii** C to G.

b The combination of these transformations can be mapped in one move. Describe fully the single transformation that maps B to G.

5 **a** Describe fully the single transformation that maps

i C to A **ii** A to I.

b The combination of these transformations can be mapped in one move. Describe fully the single transformation that maps C to I.

6 Look at the questions above. Reflecting the shapes in two perpendicular lines is equivalent to what single transformation?

S3.4HW Describing symmetries

1 Sketch as many different shapes as possible using four cubes.

Examples:

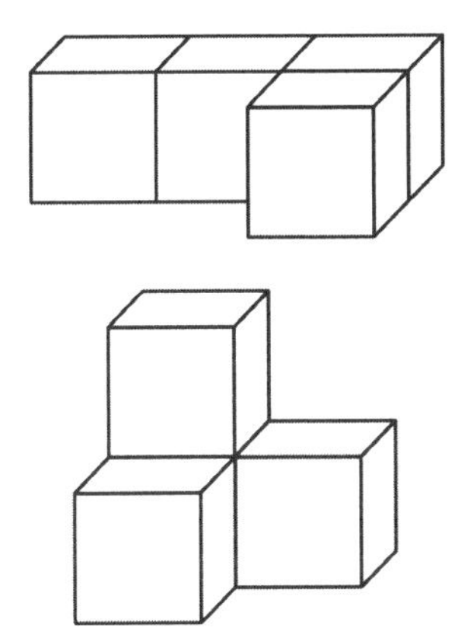

2 Explain why there are only eight different possible shapes.

3 For each shape identify any planes of symmetry.

Investigate asymmetrical shapes made from four cubes.
Look for any that form a symmetrical pair.

4 Choose one of the eight shapes.
Sketch at least four different cross-sections of the shape, showing clearly how the shape can be sliced to create them.

Copy this diagram of shape T on a grid.

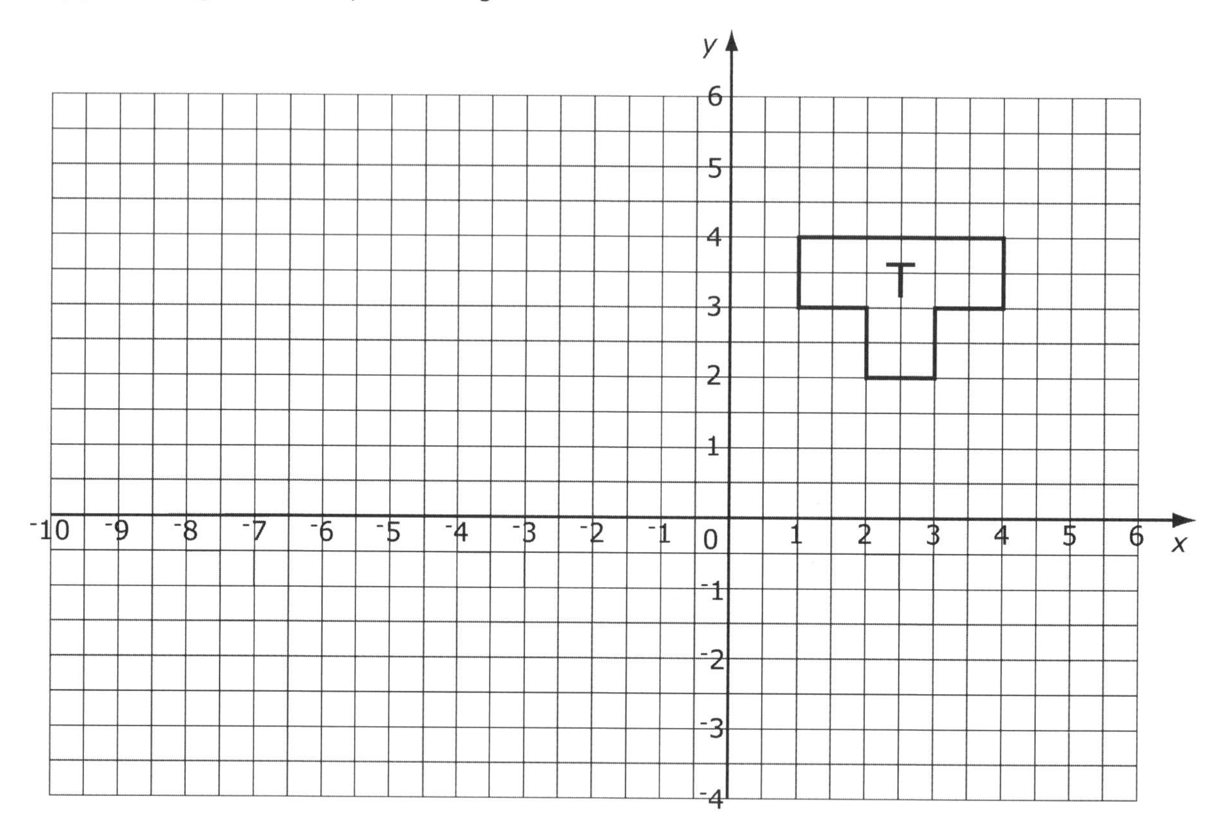

Construct the following enlargements:

a $T \to T'$: centre (2, 2), scale factor $^{-}1$.

b $T \to T''$: centre (0, 2), scale factor $^{-}2$.

What enlargements map:

c T' to T

d T'' to T?

S3.6HW Enlargement and ratio

1 Start with the shaded shape and describe a series of transformations to fill the given area. Give **precise** details.

> **Hint:**
> It may help to label the vertices of the shaded shape.

a **b** **c** **d** **e**

2 P′Q′R′S′ is an enlargement of PQRS.

a Find the centre of enlargement and the scale factor.

b What is the ratio of the perimeter of PQRS to the perimeter of P′Q′R′S′?

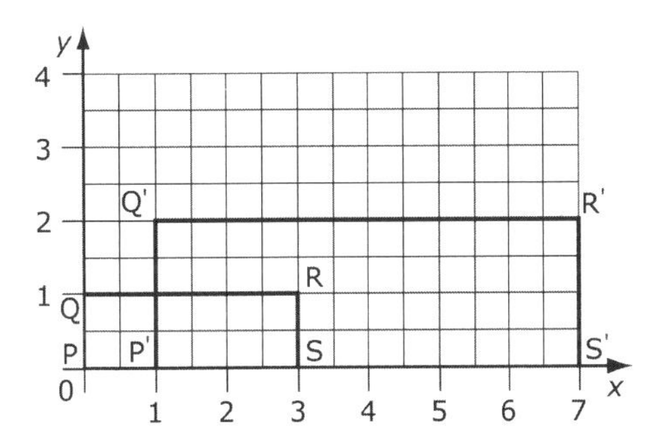

3 The ratio photo: enlargement is 2 : 9.

a What are the dimensions of the enlargement?

b What is the ratio of the areas?

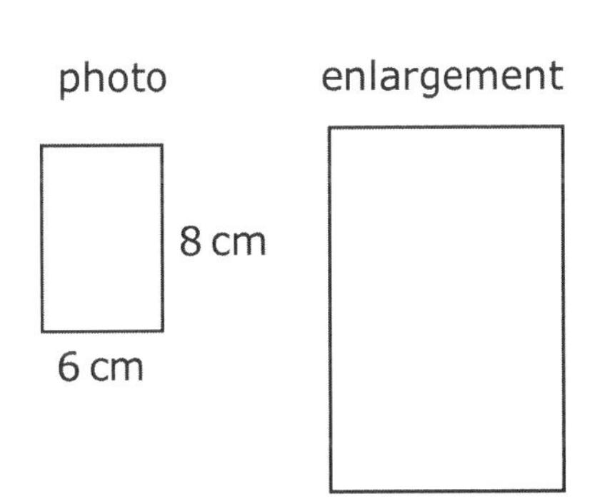

4 A logo is enlarged for a poster using the ratio 3 : 7.

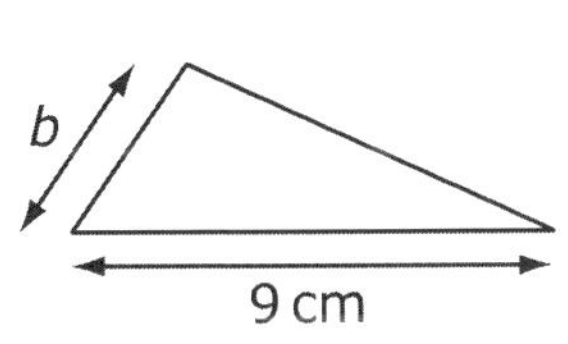

14 cm

a

Calculate *a* and *b*.

Level 6

The grid shows an arrow.

Copy the grid, and draw an **enlargement** of **scale factor 2** of the arrow.

Use **point C** as the centre of enlargement.

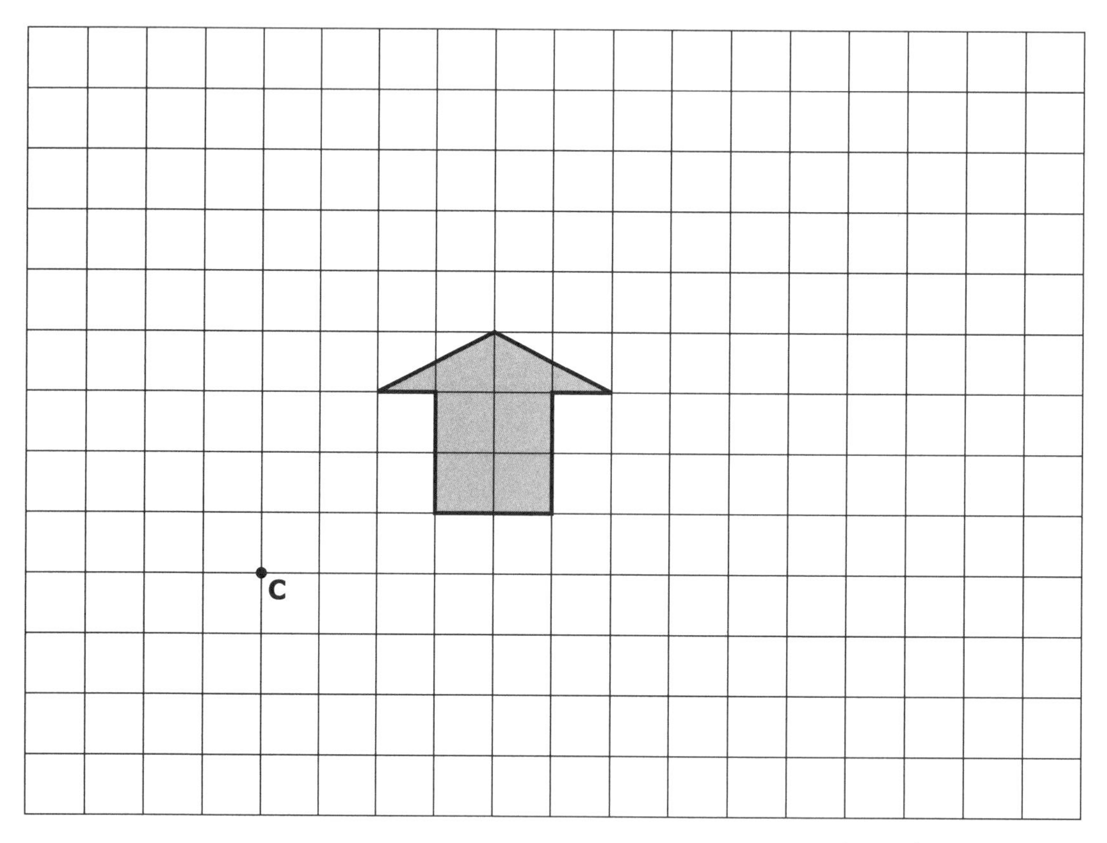

2 marks

Level 7

ill has drawn an original picture of a giraffe for an animal charity.
t measures 6.5cm high by 4cm wide.

ORIGINAL
PICTURE

Different-sized copies of the original picture can be made to **just** fit
nto various shapes.

a Jill wants to enlarge the original picture so that it **just** fits inside a rectangle on a carrier bag.
The rectangle measures 24cm high by 12cm wide.

By what scale factor should she multiply the original picture?

Show your working. *2 marks*

continued

b Jill wants to multiply **the original picture** by a scale factor so that it **just** fits inside the square shown below for a badge.

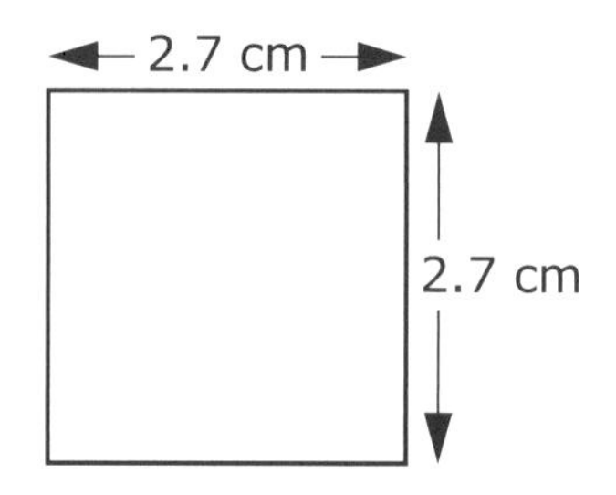

By what scale factor should she multiply the original picture?

Show your working. *2 marks*

c **The original picture** is to be used on a poster.
It must just fit inside a shape like this.

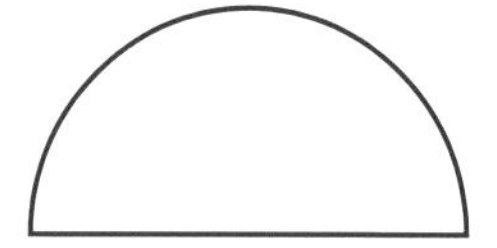

The shape is to be a semi-circle of radius 6.6cm.
What would be the perimeter of the shape?

Show your working. *2 marks*

A4.1HW Solving linear equations

1 Find pairs of equations on these cards with equal solutions:

$5x - 8 = 2x + 4$

$\frac{45}{3x} = 5$

$2(x - 3) = 5(10 - x)$

$\frac{2x + 4}{2} + 1 = 6$

$15 - 2x = 7$

$x^2 = 64$

2 Copy and complete this crossword:

1	■	■	2
3	4	■	
■	5	6	■
7	■	8	

Across

1 $x^3 = 64$

3 $50 - 2x = 8$

5 $30 - x = 74 - 3x$

7 $\sqrt{x} = 3$

8 $\frac{(2x+10)}{2} = \frac{(3x+15)}{3}$

Down

1 $2x + 18 = 102$

2 $\frac{30}{x} + 7 = 10$

4 $3(x - 9) = 2x - 15$

6 $\frac{100}{x} = 4$

3 Solve these equations by reading the layers and reversing them.
The first one is done for you.

a $2x^3 - 4 = 50$ $\quad x^3 = \frac{(50+4)}{2} = 27$ $\quad x = 3$

b $\frac{x^2}{2} + 3 = 75$

c $5\sqrt{x} - 5 = 60$

d $\frac{2(x+3)^2 \pm 20}{2} - 10 = 80$

4 Find two solutions to each of these equations:

a $x^2 = 36$ **b** $2x^2 - 8 = 90$ **c** $\frac{50}{x^2} = 2$ **d** $10 - \frac{9}{x^2} = 1$

A4.2HW Algebraic fractions

1 Simplify these expressions:

a $\frac{3}{p} + \frac{5}{p}$ **b** $\frac{k}{4} - \frac{k}{7}$ **c** $\frac{1}{2}z + \frac{1}{5}z$

d $\frac{a}{6} - \frac{b}{4}$ **e** $\frac{7}{x} + \frac{3}{y}$ **f** $\frac{10}{x} + \frac{11}{x^2}$

2 Find the missing lengths in this diagram:

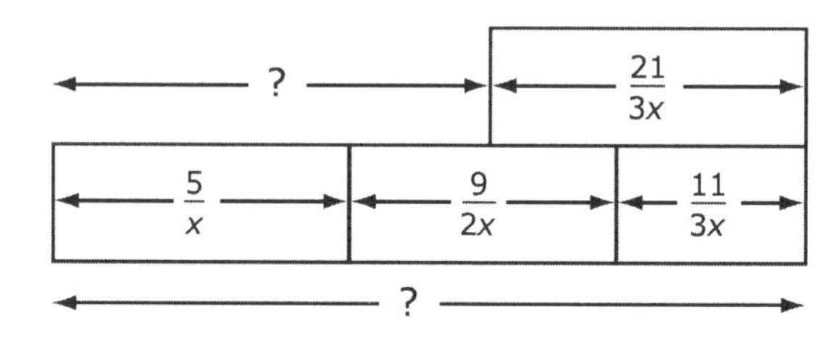

3 Spot the error in each piece of working.

a

$$\frac{a}{4} + \frac{b}{5} = \frac{5a}{20} + \frac{4b}{20} = \frac{5a + 4b}{40}$$

b

$$\frac{x}{3} + \frac{y}{7} = \frac{7x}{21} + \frac{3y}{21} = \frac{21xy}{21}$$

c

$$\frac{4}{p} + \frac{5}{p^2} = \frac{8}{p^2} + \frac{5}{p^2} = \frac{13}{p^2}$$

4 Solve these equations by adding or subtracting the fractions first:

a $\frac{10}{y} + \frac{12}{y} = 11$ **b** $\frac{11}{2x} - \frac{2}{x} = 10$ **c** $5 - \left(\frac{2}{p} + \frac{4}{p}\right) = 3$

A4.3HW Solving problems with equations

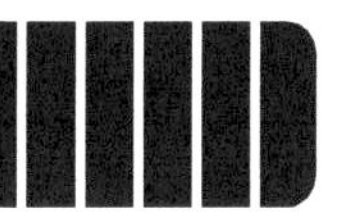

Write and solve an equation to represent these situations:

a Double a number subtracted from 10 is the same as five times the number added to 20.

b Fifteen divided by a number is equal to 23.

c When treble a number is subtracted from 50, the answer is $^{-}10$.

d A number, add two, multiplied by eight, is equal to one quarter.

The area of the shaded region is 26 cm^2.
Find the dimensions of the outer rectangle:

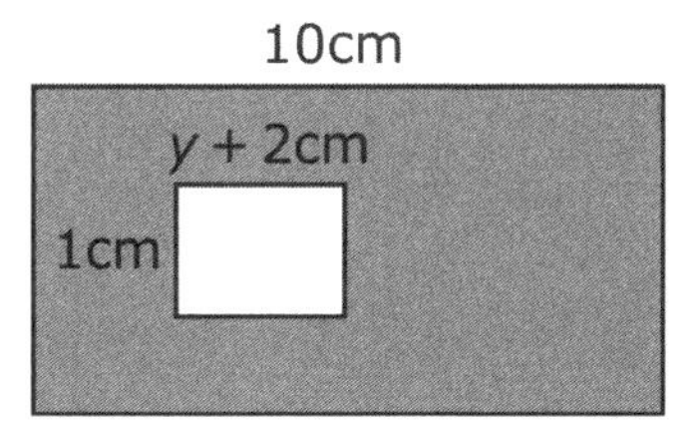

3 Peter's gym is midway between his home and work.
How far does he travel to work each day?

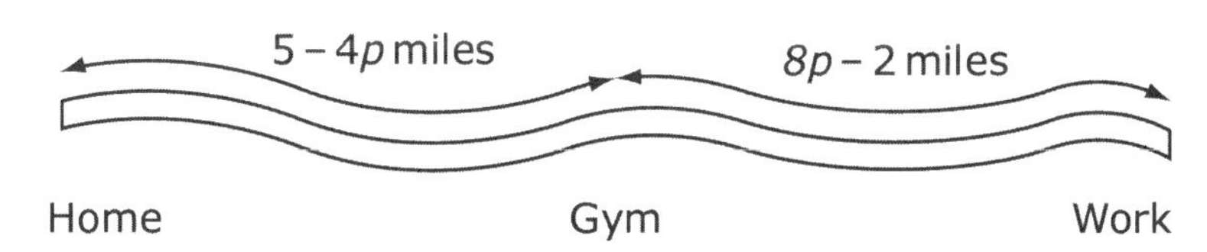

4 The areas of these shapes are equal.
What is this area in cm^2?

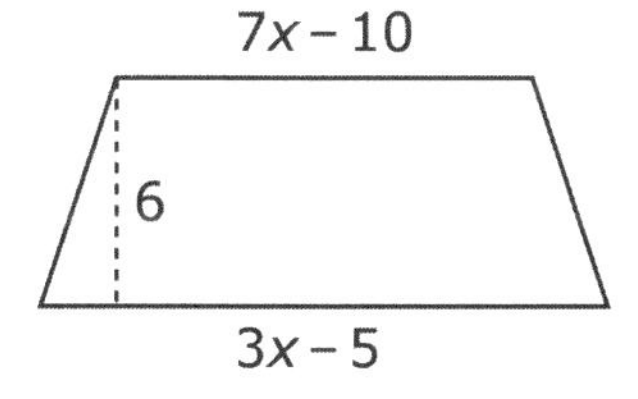

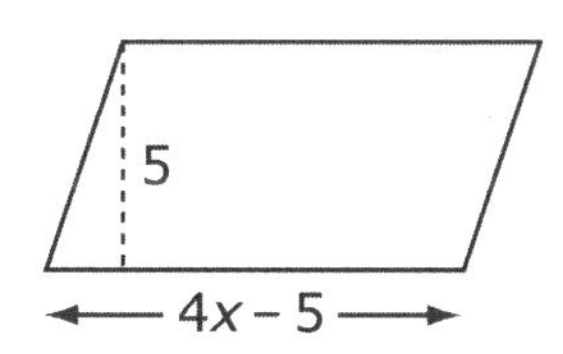

5 In ten years times I will be twice as old as I was 10 years ago.
How old and I?

A4.4HW Introducing inequalities

1 Write down if these inequalities are true or false:

a $2^3 > 3^3$ **b** $(-4)^2 < 2^2$ **c** $25^{\frac{1}{2}} < 36^{\frac{1}{2}} < 49^{\frac{1}{2}}$

d $0.3 \times 0.3 > 0.8$ **e** 11% of 340 < 10% or 370

2 Look at these number lines:

p

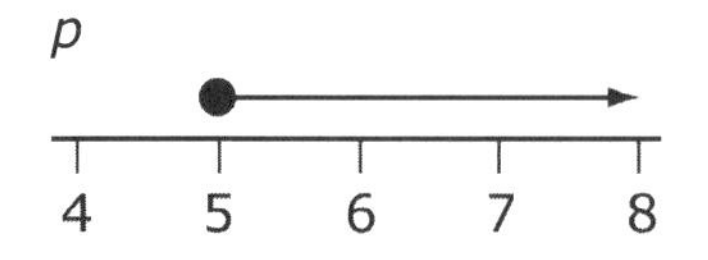

q

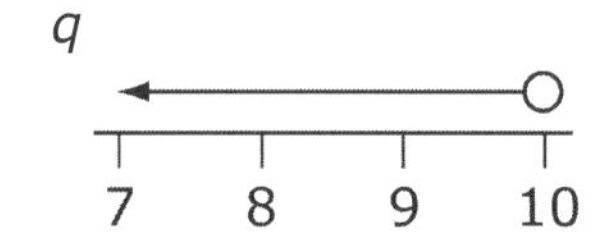

r

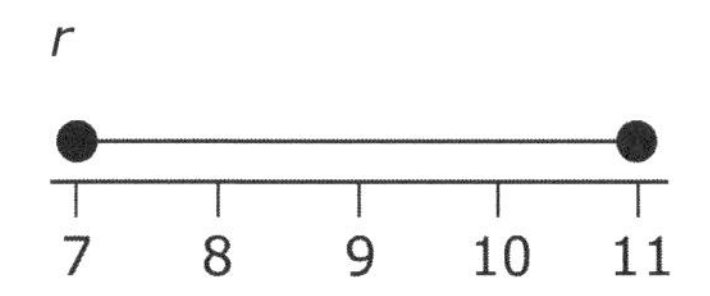

s

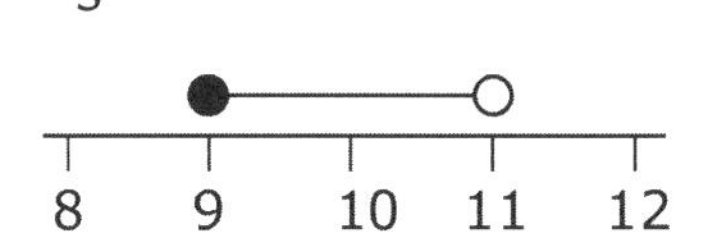

a Which numbers could have a value of 10?

b Which numbers could have a value of 3?

c If *p* and *r* are integers, what is the smallest value of the product *pr*?

d If *s* and *r* are integers, what is the biggest possible difference between them?

3 Solve these questions, representing your solutions on a number line:

a $5x \leq 40$ **b** $10y + 2 > 12$ **c** $\frac{z}{4} + 3 < 7$

d $10 - 2x \geq 4$ **e** $5y + 3 \geq 3y - 4$ **f** $2(Z - 1) + 3(Z + 4) < 8(Z + 5)$

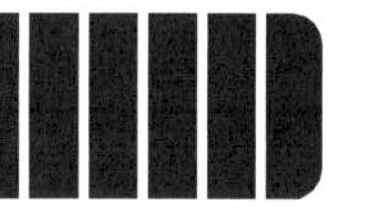

Rearranging formulae

1 A commonly used formula in physics is $\boldsymbol{v = u + at}$.
Which combination below gives the largest value of v?

a $u = 6, a = 4, t = {}^{-}2$

b $u = 2, a = 6, t = \frac{1}{2}$

c $u = 3, a = {}^{-}2, t = {}^{-}4$

d $u = 7, a = {}^{-}2, t = {}^{-}3$

2 The approximate surface area of a sphere can be found using the formula:

$$S = \frac{88r^2}{7}$$

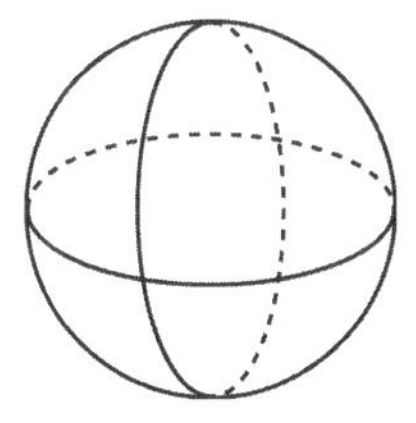

a Find the approximate surface area of a sphere with radius 6 cm to two decimal places.

b If a sphere has a surface area of 100 cm^2, what is its radius to one decimal place?

3 Rearrange these formulae to make x the subject:

a $px + q = w$ **b** $wx - z = r$ **c** $\frac{x+p}{k} = z$

d $x^3 + k = p$ **e** $\frac{x}{p} - k = w$ **f** $\sqrt{x} + b = k$

4 The formula $F = \frac{9C}{5} + 32$ can be used to convert a temperature from degrees Centigrade (C) to degrees Fahrenheit (F).

a Use the formula to find the freezing and boiling point of water in degrees Fahrenheit.

b Rearrange the formula to make C the subject.

c Use your answers from **b** to find 50 °F in °C.

d Which is greater, 30 °C or 80 °F?

A4.6HW Rearranging harder formulae

1 Rearrange these equations to make p the subject:

a $\frac{pw + k}{q} - z = m$ **b** $\frac{m(kp^2 - r)}{n} = w$ **c** $\frac{k(p^3 + r) - z}{m} = n$

2 Rearrange these equations to make a the subject.
Take extra care with negative and fractional terms.

a $p - a = k$ **b** $q - ma = w$ **c** $\frac{p \pm a^2}{t} = r$

d $r - \sqrt{a} = z$ **e** $\frac{m}{a} = z$ **f** $\frac{t}{pa} = m$

g $\frac{t}{a} - k = w$ **h** $k = m - \frac{p}{a}$

3 Norman pays for some books with a £50 note.
Each books costs £2.

a Copy and complete this table:

No. of books b	1	2	3	4	5
Change from £50, C					

b Find a formula for C, the change from £50, in terms of b, the number of books that Norman buys.

c Rearrange the formula to make b the subject.

4 A cyclist travels 50 miles at a speed of 26 mph.
A boy walks 6 miles at a speed of 3.5 mph.

Whose journey takes longer?

> **Remember:**
> $\text{Speed} = \frac{\text{Distance}}{\text{Time}}$

Level 6

Look at these equations.

$3a + 6b = 24$ $2c - d = 3$

a Use these equations to work out the value of the expressions below.

The first one is done for you.

i $8c - 4d = 12$

ii $a + 2b =$

iii $d - 2c =$ *2 marks*

b Use one or both of the equations to write an expression that has a value of **21**. *1 mark*

Level 7

a The subject of the equation below is p:

$$p = 2(e + f)$$

Rearrange the equation to make e the subject. *2 marks*

b Rearrange the equation $r = \frac{1}{2}(c - d)$ to make d the subject.

Show your working. *2 marks*

D2.1HW Planning data collection

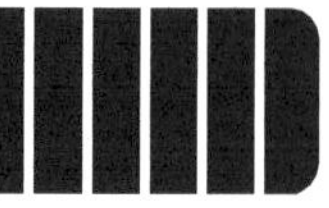

Robert and Rose are completing jigsaw puzzles.

Rose thinks that girls are quicker than boys at completing the puzzles.

Robert believes that boys are better than girls at puzzles because he had just completed his puzzle with the pieces face down (so that the picture could not be seen).

Robert and Rose decide to investigate how long it takes boys and girls to complete jigsaw puzzles.

1 What sort of data will they need to collect?

2 Identify possible sources of the data.

3 Write a detailed plan of how they might collect the data and how they could record the data.

Displaying data

Remember:

A back-to-back stem-and-leaf diagram looks like this:

Team 1					Team 2			
			2	3	6	6	7	
5	2	1	1	2	2	4	5	
9	6	5	5	1	3	3	6	9
	4	3	2	0	1	4		

Where 2|3|6 means 32 for Team 1 and 36 for Team 2.

Rose collected data on how long it took pupils in year 8 to complete a simple jigsaw puzzle.

The tables below give her results.

Boys' times (seconds)		
64	33	42
29	48	51
34	58	62
56	61	60
49	63	52

Girls' times (seconds)		
58	75	61
44	52	60
48	58	65
48	44	40
47	46	52

Draw a back-to-back stem-and-leaf diagram to represent this data.

D2.3HW Interpreting data

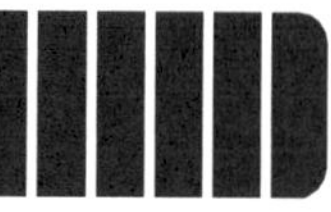

Robert collected data on how long, in seconds, it took students in year 8 to complete a simple jigsaw with the picture face down.

He drew a back-to-back stem-and-leaf diagram to present his data.

		Boys								Girls		
				2	10							
	7	4	3	1	9	2	4					
8	6	5	5	3	8	1	3	4	4	6	8	9
	9	6	3	0	7	0	4	5	7	9		
		6	5	5	6	7	8	8				

Key: 1 | 9 | 2 means 91 seconds for boys and 92 seconds for girls

1 Find the mean, median and range for both sets of data.

2 Use your results and the shape of the diagram to comment on the time it takes boys and girls to complete a jigsaw with the picture face down.

3 Draw frequency diagrams to illustrate the data.

D2.4HW Scatter graphs

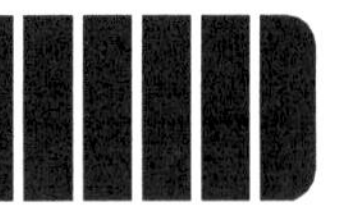

Robert asked all the boys in his survey to complete the jigsaw with the picture face down and then face up.

The table shows his results.

All times are given in seconds.

Picture face down	65	65	66	70	73	76	79	83	85	85	86	88	91	93	94	97	102
Picture face up	42	49	46	48	52	54	55	60	61	62	63	64	64	64	66	65	69

1 Draw a scatter graph to display this data.

2 Comment on any trend shown by your graph.

3 What type of correlation does it show?

4 Another boy completes the jigsaw with the picture face down in 68 seconds.
Use your graph to suggest what his time might be for the picture face up.
Justify your answer.

D2.5HW Comparing data

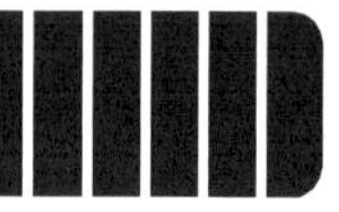

Here are the times taken for a group of boys and a group of girls to complete a simple jigsaw.

Boys' times (seconds)			Girls' times (seconds)		
64	33	42	58	75	61
29	48	51	44	52	60
34	58	62	48	58	65
56	61	60	48	44	40
49	63	52	47	46	52

1 Explain why the mode, median and range are not the best measures to use to compare these sets of data.

2 Calculate the mean for each set of data.

3 Comment on the mean values you found and whether you think they are representative of the data for the boys and girls respectively.

4 **a** Another boy is included in the survey and the mean changes to 51.3 s.
What was his time, to the nearest second?

b Another girl is included in the survey and the mean changes to 52 s.
What was her time, to the nearest second?

D2.6HW Statistical reports

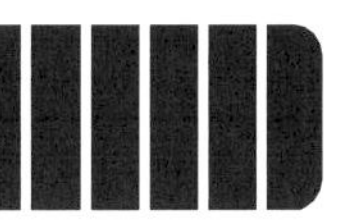

Congestion charges for London motorists were introduced in February half term, 2003.

The table shows the journey times, to the nearest minute, during February half term in 2002 and 2003 for 12 key routes in London.

The times are based on information from Trafficmaster's network of cameras.

	2002	2003
A10 Waltham Cross to Stoke Newington	44	38
A12 Harold Wood to Blackwall Tunnel	73	68
A2/A102 Dartford to Blackwall Tunnel	44	33
A20 Swanley to Eltham	17	16
A23 Hooley to Brixton	55	47
A3 Cobham to Clapham	42	46
A316 Sunbury Cross to Ravenscroft	26	27
A30 Stanwell to Osterley	13	13
A4 Langley, Slough to Talgarth Road	46	46
A40(M) Denham to Marylebone	54	39
A41 Mill Hill, Fiveways to Regent's Park	27	19
A1 Mill Hill to Islington	51	28

1 Draw a scatter graph to represent this data.

2 Calculate appropriate statistics to represent this data.

3 Write a short report on the effect of the introduction of congestion charging in London in February 2003. Refer to your statistics and your graph.

Level 6

A teacher asked two different classes:

'What type of book is your favourite?'

a Results from **class A** (total 20 pupils):

Type of book	Frequency
Crime	3
Non-fiction	13
Fantasy	4

Copy and complete the pie chart to show this information.
Show your working and draw your angles accurately.

Class A

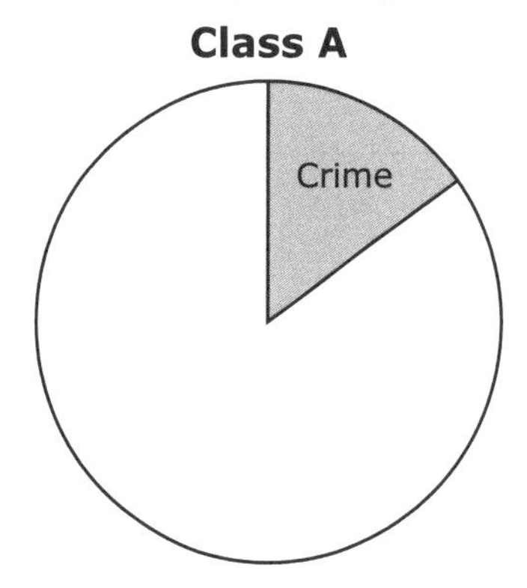

2 marks

b The pie chart below shows the results from all of **class B.**

Each pupil had only one vote.

Class B

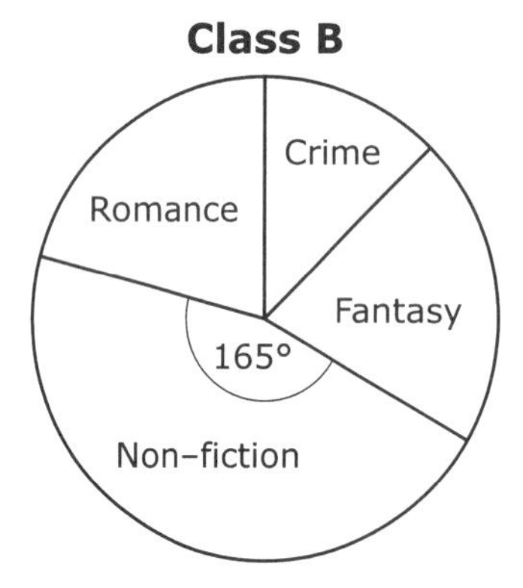

The sector for **Non-fiction** represents **11 pupils**.

How many pupils are in class B?

Show your working. *2 marks*

Level 7

From 5th May 2000 to 5th May 2001 a swimming club had the same members.

Copy and complete the table to show information about the ages of these members.

Ages of members	
Mean (5th May 2000)	24 years 3 months
Range (5th May 2000)	4 years 8 months
Mean (5th May 2001)	
Range (5th May 2001)	

1 mark

The table below shows information about members of a different club.

Ages of members	
Mean	17 years 5 months
Range	2 years 0 months

A new member, aged **18 years 5 months,** is going to join the club.

What will happen to the **mean** age of the members?

Write down the correct statement below.

- It will increase by more than 1 year.
- It will increase by exactly 1 year.
- It will increase by less than 1 year.
- It will stay the same.
- It is not possible to tell. *1 mark*

What will happen to the **range** of ages of the members?

- It will increase by more than 1 year.
- It will increase by exactly 1 year.
- It will increase by less than 1 year.
- It will stay the same.
- It is not possible to tell. *1 mark*

Addition and subtraction lines

1 This number cloud shows fractions and decimals.

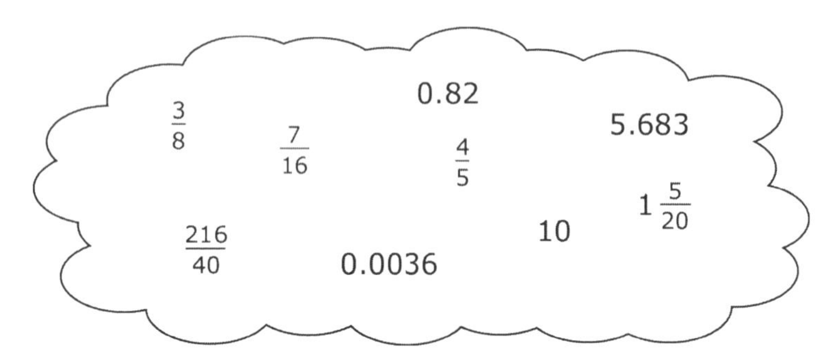

a Work out an estimate for the sum of the numbers in the cloud.

b Work out the exact sum of the numbers.

c Calculate the difference between the smallest number and the largest number.

2 Use a standard written method to calculate the sum and difference of 6.72 and 20.3082.

3 In an addition pyramid, each brick is the sum of the two bricks beneath it. For example:

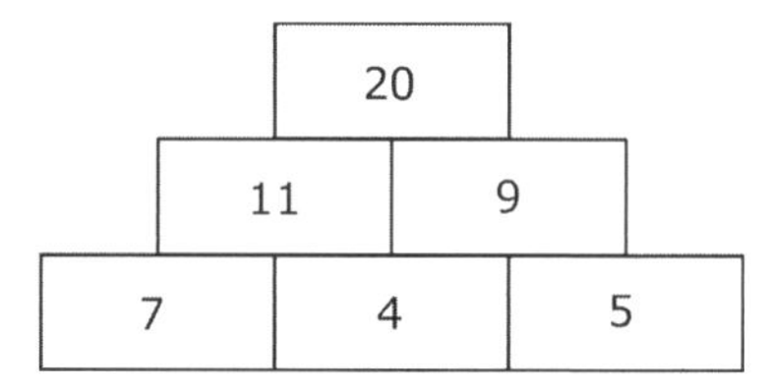

Copy and complete these pyramids.

a

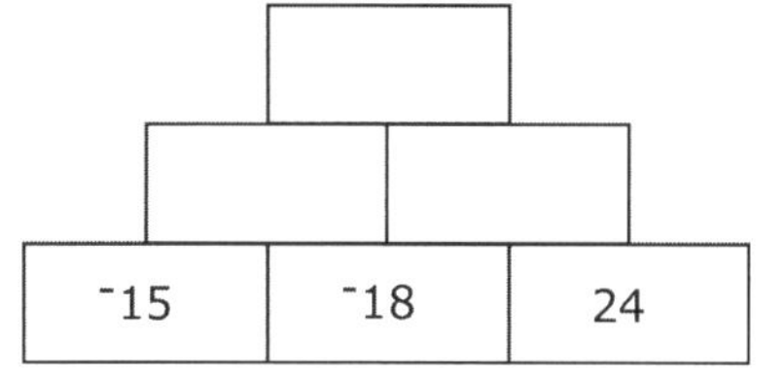

b

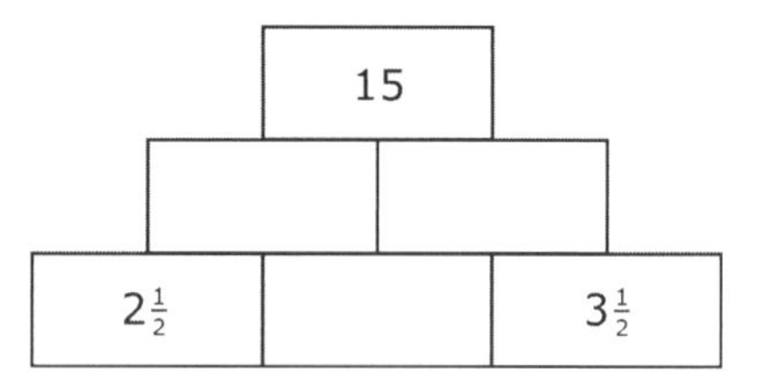

c

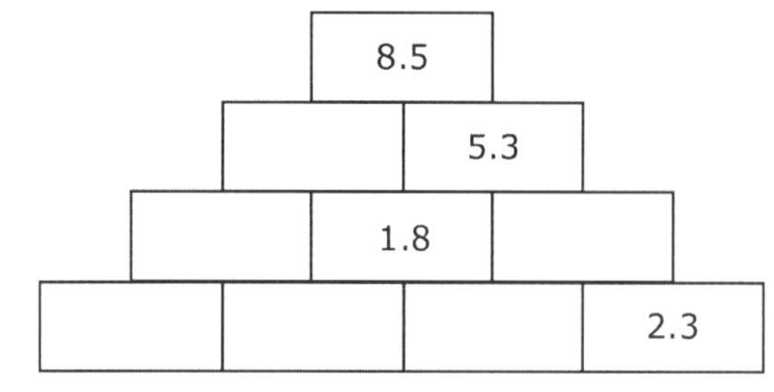

Multiplication and division laws

Calculate these:

a $^-26 \times 0.4$ **b** $51 \div \frac{1}{4}$

c $\frac{3}{11} \times {}^-4$ **d** $^-\frac{1}{6} \div 3$

e $^-\frac{1}{5} \times 18$ **f** $^-4.5 \div 2$

g $\frac{2}{7} \times 18$ **h** 4.8×15

2 **a** Work out the value of each of these expressions if:

$a = 7$ $b = 0.4$ $c = {}^-0.3$ $d = {}^-8$

Write your answers as fractions and decimals.

i $c \div a$ **ii** cad

iii $3b \div 2d$ **iv** c^2

3 Copy and complete these Multiplication Pyramids.
The number in each cell is the product of the two below it:

a

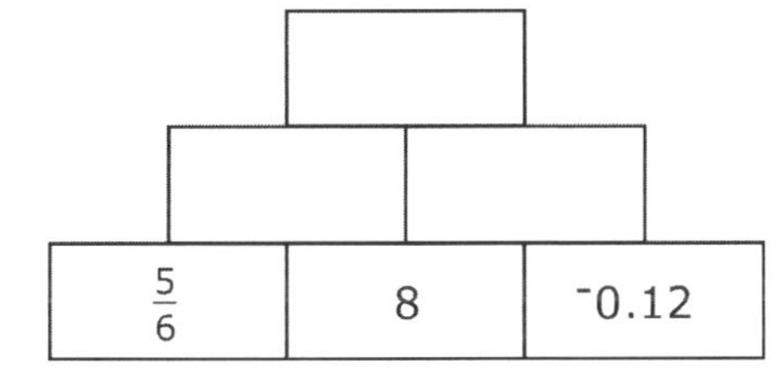

b

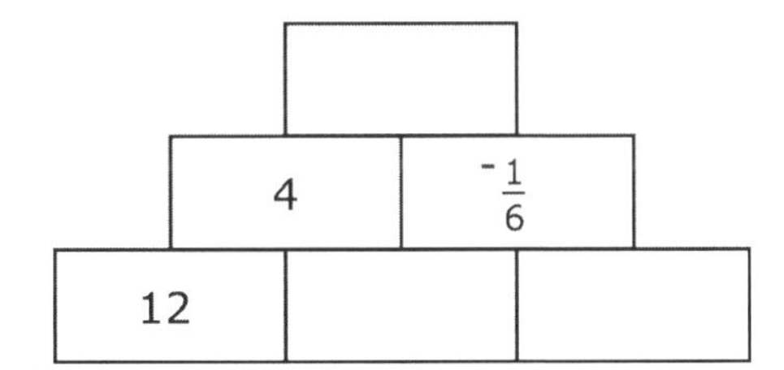

N4.3HW Order of operations

Remember:
The order of operations:

- **B**rackets
- **I**ndices
- **D**ivision and **M**ultiplication
- **A**ddition and **S**ubtraction

1 Work out each of these using a calculator. Estimate first.
Give your answers to 2 d.p.

a $(5.3)^3 - 2.7 \div 3.14$

b $(6.9 - 4.02)^2 \div 6$

c $4\frac{1}{8} \times (6.3 - 2.5)$

d $\dfrac{(8.3 - 3.1)^3}{8}$

e $9 + \frac{3}{4} \div 3.6$

f $\dfrac{2.1}{5.6^3} - 9.4^2$

2 Use your knowledge of inverse operation to work out each of these.

a If $p^2 = 110$, find the value of p correct to 2 d.p.

b If $a^4 + 6 = 1000$, find the value of a correct to 1 d.p.

c The volume of this cube is 2500 cm^3.

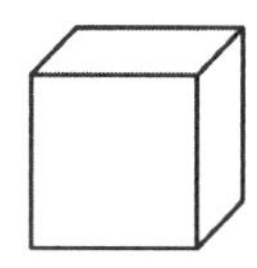

Use this information to work out:

i The side length of the cube (write down the full calculator display).

ii The area of one face of the cube correct to 1 d.p.

N4.4HW Multiplying decimals

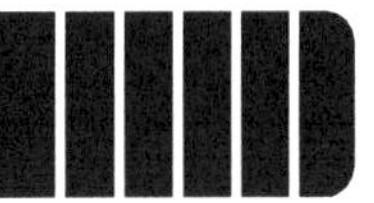

1 Puzzle

Calculate the missing numbers in the puzzle. The numbers in the diamonds are the products of the numbers in the circles on either side.

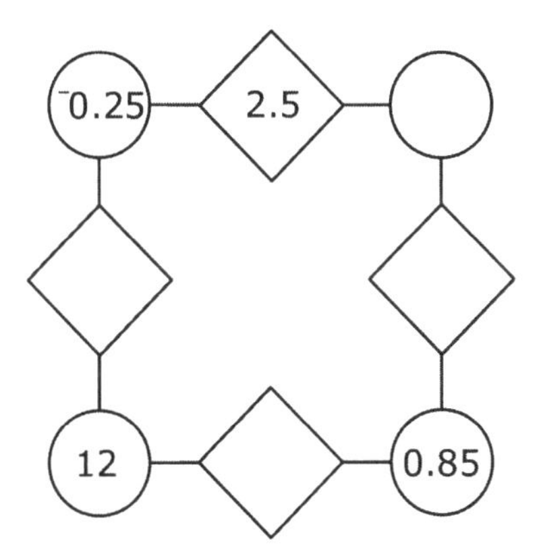

2 Use a written method to work out the exact answer to each of the following. You must write an approximate answer before you complete each calculation.

a 23.1 x 5.2

b 0.52 x 7.9

c 6.96 x 3.2

d 1.34 x 5.39

N4.5HW Dividing decimals

1 Calculate the missing values giving your answers to 2 d.p. where appropriate:

a $3.41 \div \square = 5.5$

b $8 \div \square = 31.2$

c $7.5 \div \square = 0.312$

d $\square \div 0.65 = 0.0678$

e $\square \div 0.23 = 7.52$

f $\square \div 230 = 4.351$

2 Express the missing value as a fraction (or mixed number if appropriate) in its simplest form:

a $\square \div {}^{-}\frac{2}{5} = \frac{4}{15}$

b $\frac{3}{5} \div \square = \frac{7}{55}$

c $\frac{5}{12} \div \square = {}^{-}\frac{9}{84}$

d $\square \div \frac{3}{4} = 8\frac{17}{32}$

e ${}^{-}4\frac{7}{10} \div \square = 3\frac{17}{40}$

f $\square \div {}^{-}3\frac{1}{2} = {}^{-}18\frac{2}{5}$

Design a poster

Design a poster (A4) to show why:

100 cm = 1 m

but

100 cm^2 ≠ 1 m^2

You can include:

- Rectangular diagrams to show area.
- Calculations of areas of different sized shapes, for example, 10 cm by 10 cm, 100 cm by 100 cm, 1 m by 1 m.

Extend the idea to show the connection between cm^2 and mm^2.

- You may want to consider the areas of rectangles of 1 cm by 1 cm, 10 mm by 10 mm and 1 mm by 1 mm.

Finish by completing this statement:

1 cm^2 = ___ mm^2

Level 6

A company sells and processes films of two different sizes.

The tables show how much the company charges.

Film size: **24** photos	
Cost to **buy** each film	£2.15
Postage	free
Cost to **print** each film	£0.99
Postage for each film	60p

Film size: **36** photos	
Cost to **buy** each film	£2.65
Postage	free
Cost to **print** each film	£2.89
Postage for each film	60p

I want to take **360** photos.

I need to buy the film, pay for the film to be printed, and pay for the postage.

Is it cheaper to use all films of size 24 photos, or all films of size 36 photos?

How much cheaper is it? Show your working. *4 marks*

Level 7

a Write down the **best** estimate of the answer to:

$72.34 \div 8.91$

6 7 8 9 10 11 *1 mark*

b Write down the **best** estimate of the answer to:

32.7×0.48

1.2 1.6 12 16 120 160 *1 mark*

c Estimate the answer to $\frac{8.62 + 22.1}{5.23}$.

Give your answer to **1 significant figure**. *1 mark*

d **Estimate** the answer to $\frac{28.6 \times 24.4}{5.67 \times 4.02}$. *1 mark*

A5.1HW Investigating a task

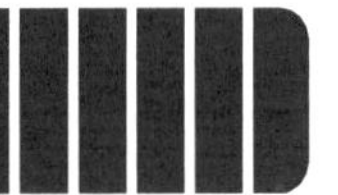

1 Find a formula for these totals by breaking the task into steps:

a The total when n odd numbers are added.

$1 + 3 + 5+ 7+ \ldots$

b The total when n even numbers are added.

$2 + 4 + 6 + 8 + \ldots$

c The total when n consecutive numbers are added.

$1 + 2 + 3 + 4 + 5 + \ldots$

2 Three supermarket workers are stacking tins in a display.
The number of layers of tins they stack depends on the number of hours they have to work.

These diagrams show their stacks after three hours:

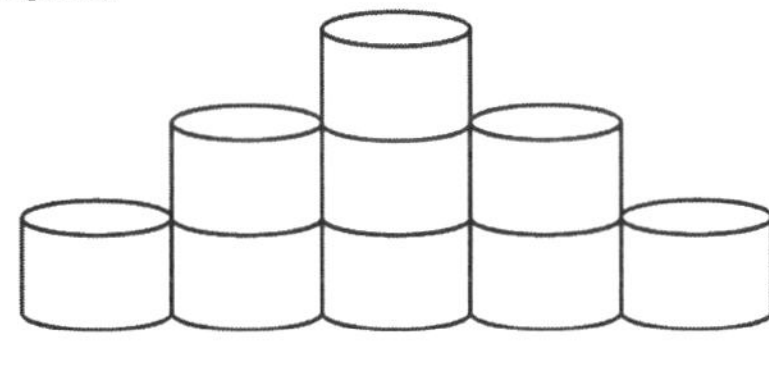

Aled

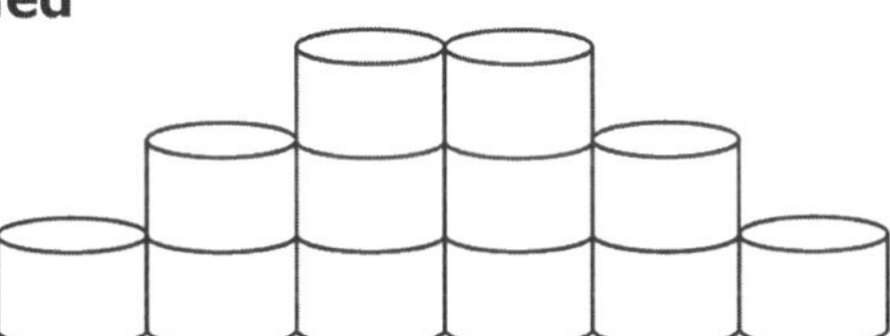

Bob

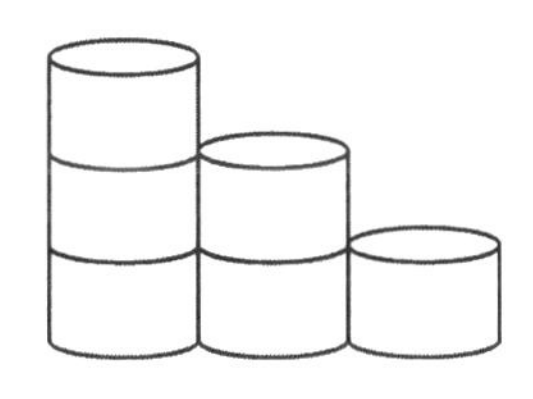

For each worker, investigate how many tins they can stack in any number of hours.

Generalise your findings into a formula.

Explain why each formula works.

Generalising findings

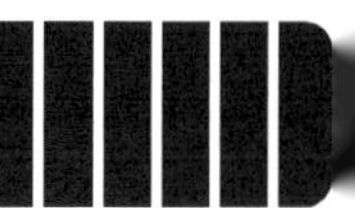

1 A pond is to be added inside a square garden of various sizes:

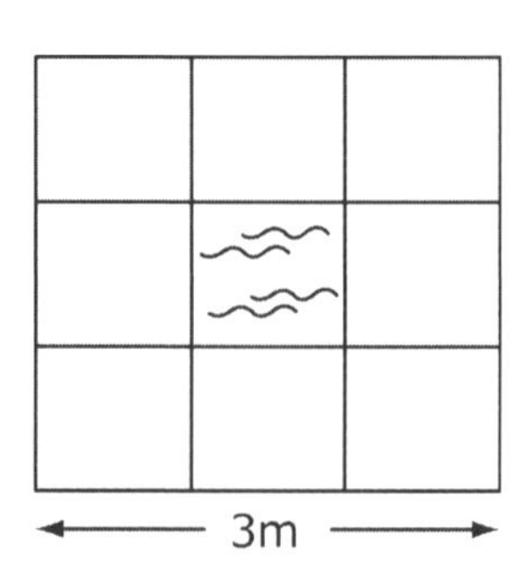

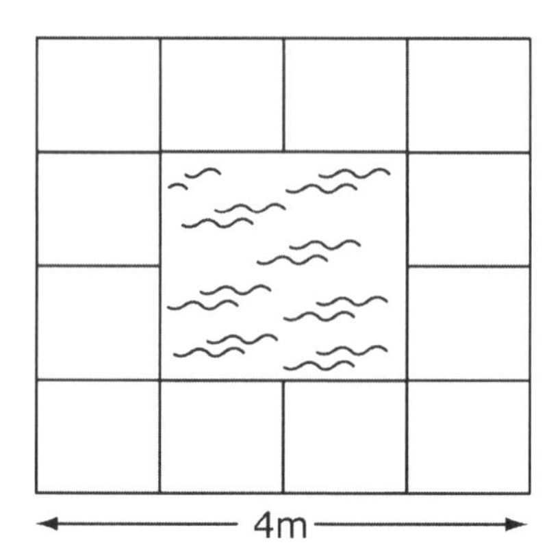

a Investigate the length of the pond in differing size square gardens.

Generalise your results into a formula.

b Investigate the area of the pond in differing size square gardens.

Generalise your results into a formula.

c Extend your investigation into rectangular ponds and gardens, and repeat parts **a** and **b**.

2 **Goal!**

a A football match ends in a draw.

Investigate, for different final scores, the number of possible scores at half-time.
For example:

A 1 – 1 draw after full time could have scores 1 – 1, 1 – 0, 0 – 1 or 0- 0 at half-time.

Generalise and explain your findings.

b Extend your investigation into a match where the final score is not a draw.

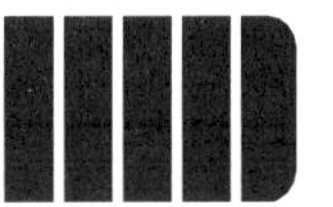

A5.3HW Factorisation

1 Factorise these expressions fully:

a $36 + 12b$ **b** $3ab - 9$ **c** $4a^2 + 16a$ **d** $5x + 10xy$

2 Show that the formula for the perimeter of this rectangle is:

$P = 12x(2y + 1)$

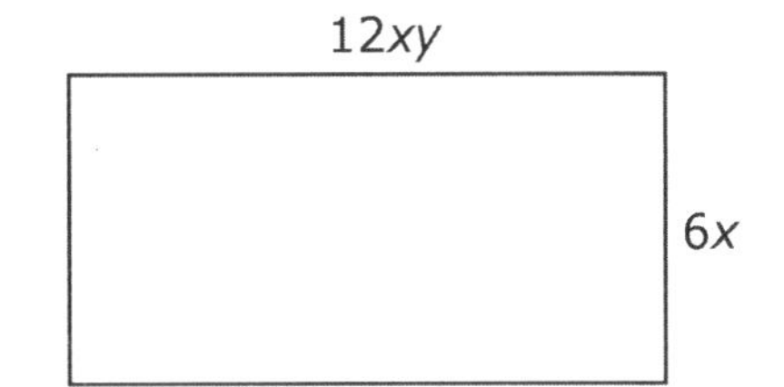

3 **a** Write down a formula for the shaded area of this rectangle.

b Write down a factorised formula.

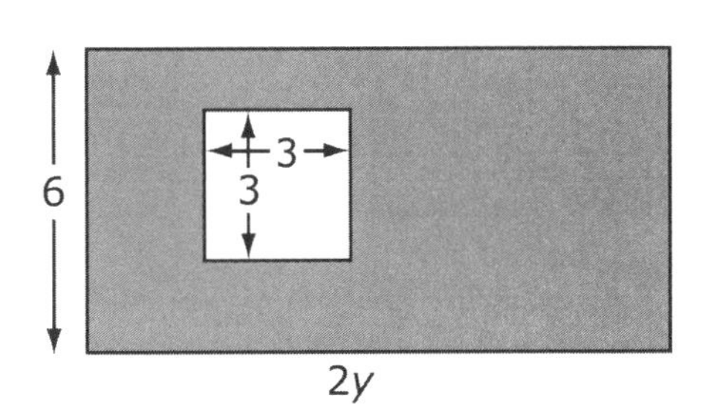

4 'Multiplying a number by 3 and adding 6 is the same as adding 2 then multiplying by 3.'

a Show that this is true with two numbers of your choice.

b Use algebra to prove that this statement is true.

c Make up and prove a similar statement of your own.

5 A formula for the number of blocks in a rectangle of height h is:

$B = h^2 + 2h$

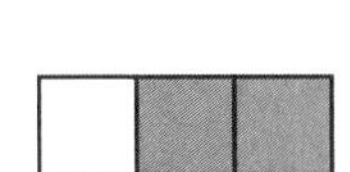

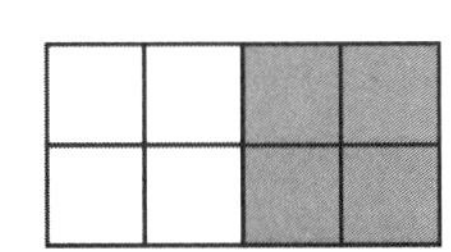

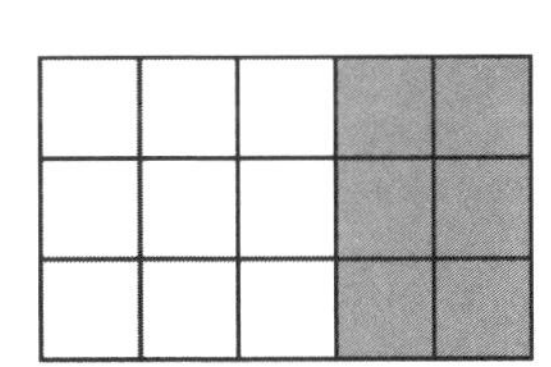

a Explain how this formula is shown by the diagram.

b Factorise this formula.

c Explain how your factorised formula is shown in the diagram

A5.4HW Trial and improvement

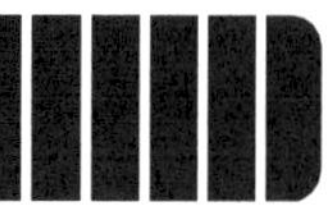

Remember:
Keep your trials **systematic.**

1 Solve these equations using trial and improvement:

a $h^2 + h = 90$ **b** $2k^2 + 1 = 20.22$

2 Solve these equations using trial and improvement, giving your answers to the correct number of decimal places.

a $b^3 - b = 90$ (to one dp).

b $m^5 - 3 = 100$ (to 2 dp).

3 a Write down a formula for the area of a rectangle whose length is 2 cm greater than its width.

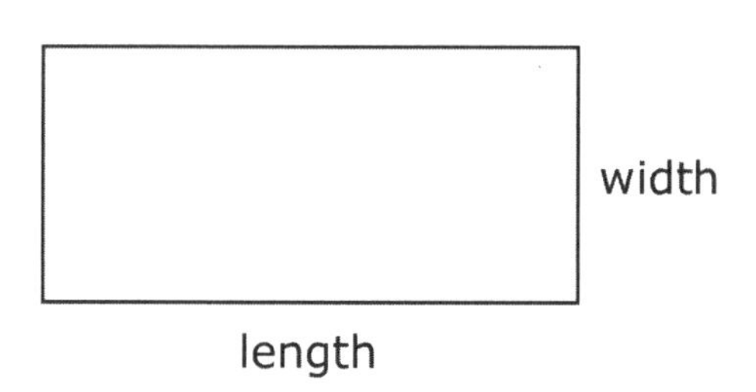

Hint: Only use one variable.

b If the area of the rectangle is 85 cm^2, find its dimensions. Give your answers to 1 decimal place.

4 When a ball is thrown in the air, its height during its flight is given by the equation:

$h = 5t - t^2$

Where h = height in metres and t = time in seconds.

Use trial and improvement to find the maximum height of the ball during a five second flight.
Give your answer to one decimal place.

Hint:
t can have any value between 0 and 5.

A5.5HW Graphs of proportions

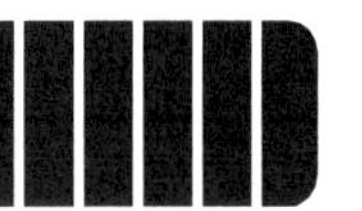

1 Which pairs of quantities are in direct proportion?

   a Number of guests for dinner – Number of plates needed.

   b Number of bags of crisps bought – Total cost of crisps.

   c Age of a car – Price of a car.

   d Weight in kg – Weight in stone.

2 In a bank, £1 can be exchanged for 1.6 Euros.

   a Draw a graph from £0 to £100 to show this conversion.

   b Use your graph to find:

      i The number of Euros you could buy for £53.

      ii The number of pounds you could buy for 130 Euros.

   c Write down the equation of your graph.

3 Here are three sketch graphs, each with their equation.
Explain what each graphs shows, in terms of the two variables given.

a

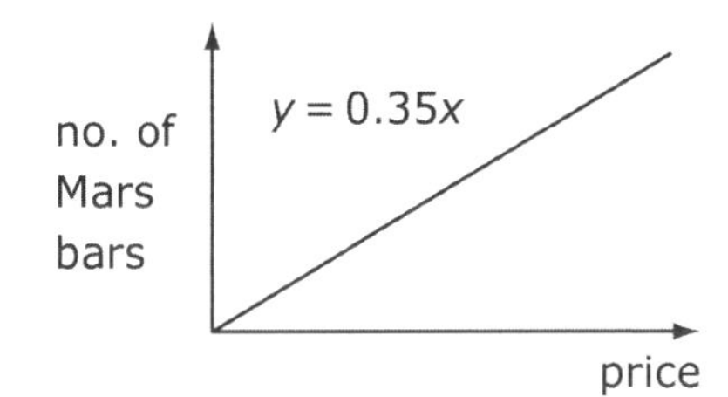

b

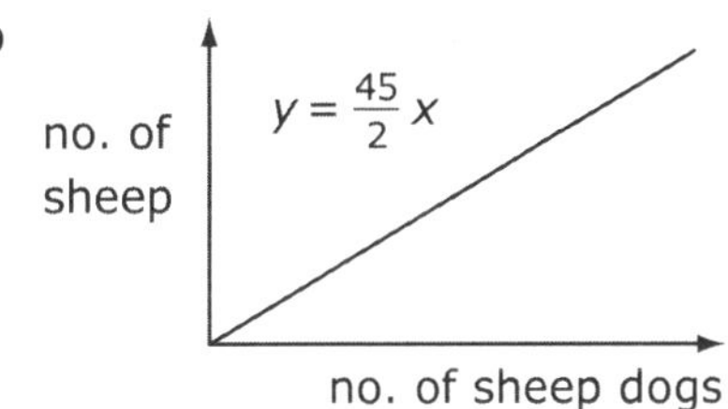

c

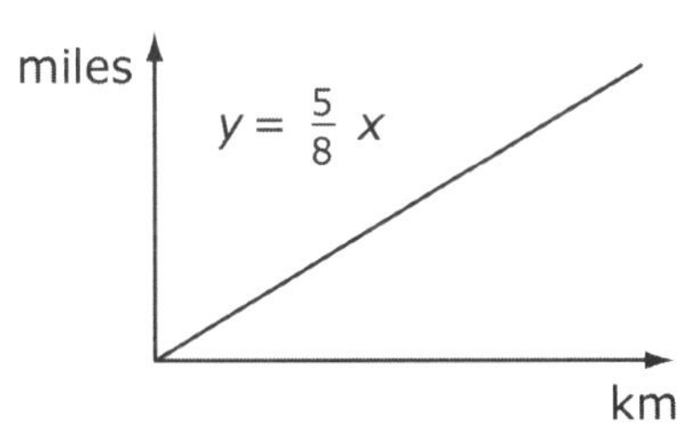

4 As the length of a square increases, so does its area.

Why are 'length of square' and 'area of square' not in direct proportion?

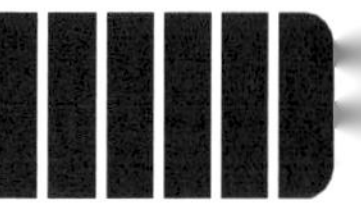

A5.6HW Proportion and algebra

1 All the quantities in this question are in direct proportion.
Find the missing value in each case, using algebra.

a

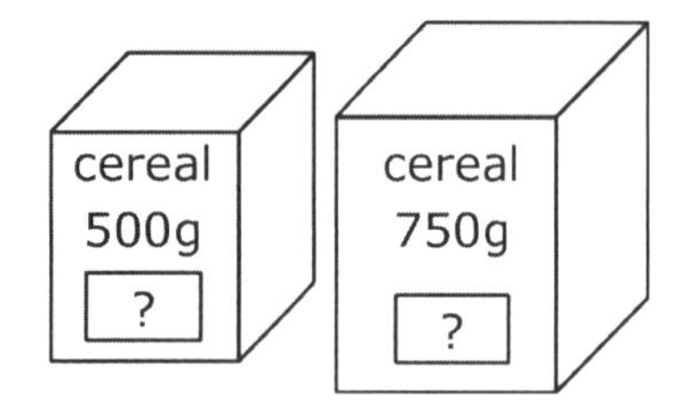

b

c A car travelling at 70 mph travels 160 miles in a certain time.
The same car travelling at 55 mph travels _____ miles.

2 At E-Z-Go courier company, the number of hours is proportional to the pay for day shift workers.

a Copy and complete this table for day shift pay.

Hours	2	5	6	7	9
Pay			£32.40		

b John gets £70 for working an 11-hour night shift.
Does the pay stay in proportion to the day shift?
Give a possible reason for your answer.

c If Sarah does three 10-hour days and two 8-hour days in one week, how much does she earn?

3 Show that, in each part, the two variables are not in proportion, and decide which is the 'best buy' in each case.

a

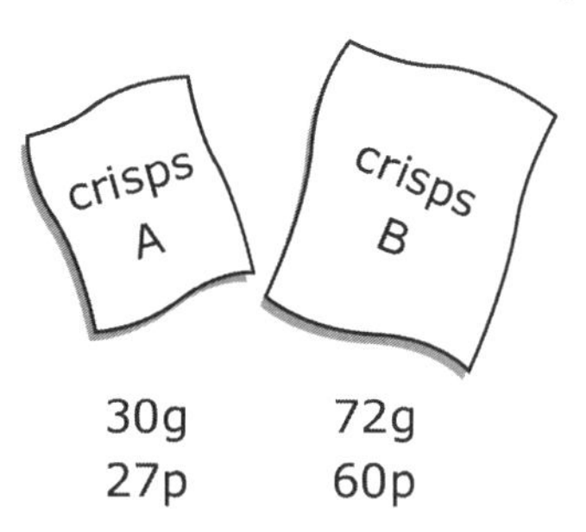

30g 27p
72g 60p

b

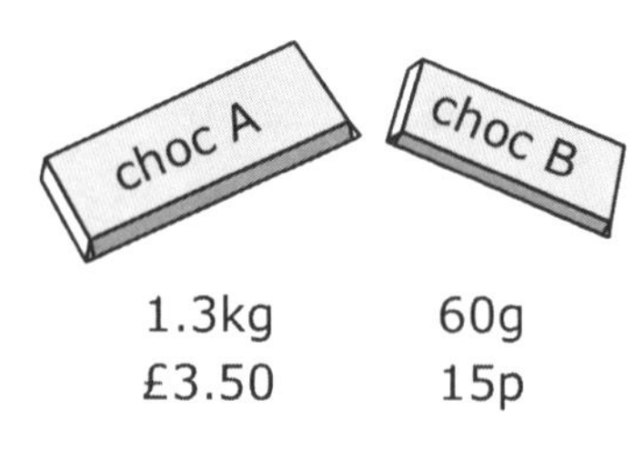

1.3kg £3.50
60g 15p

A5.7HW Real-life graphs

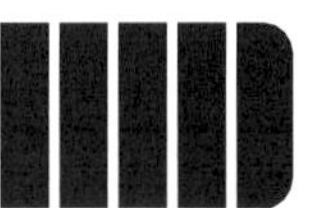

1 Sam invests some money in a savings account and leaves it for 10 years.
The graph shows the amount of money he has in the bank during the 10 years.

a How much did he invest?

b Why does he have more money after 10 years?

c How much do his savings go up by each year?

d Write an equation for the graph.

e When will his savings reach £1280?

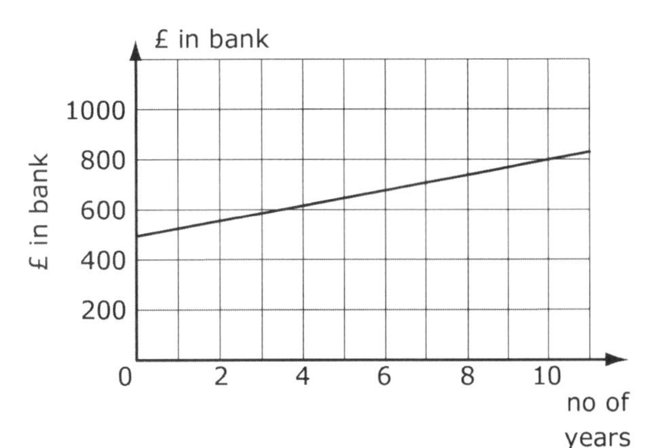

2 Write an equation for each graph, and explain what it shows.

a

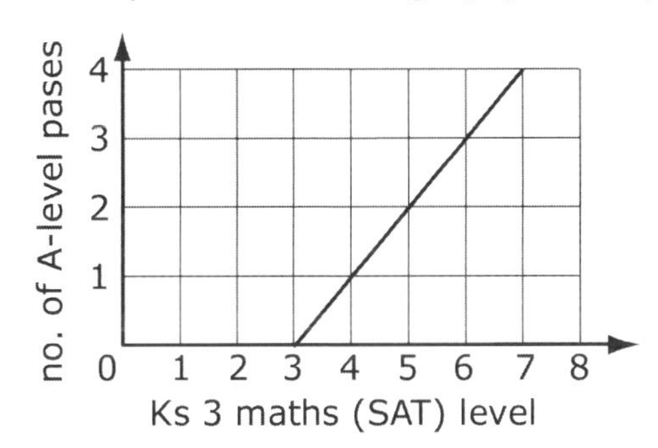

b

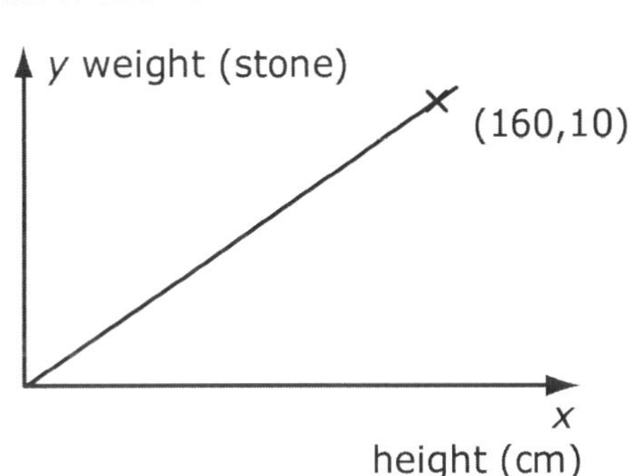

c

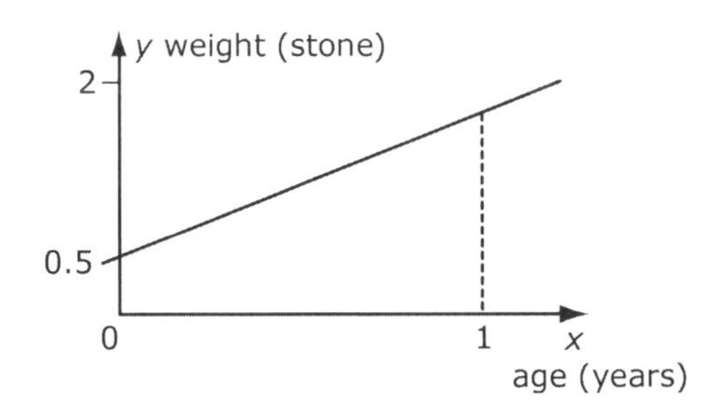

3 A hotel uses this equation to represent the cost of a one night stay at the hotel (y) given the number of room service calls (x).

$y = 5x + 40$.

Explain the hotel's charging system in words.

A5.8HW Further line graphs

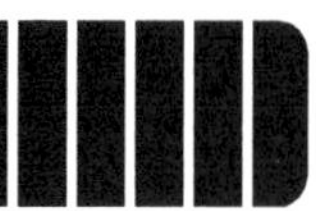

1 Plot these implicit graphs, using a table of values like this.

x	0	1	2	3	4	5
y						

a $3x + 5y = 15$ **b** $x - 2y = 10$ **c** $4x + 7y = 20$

2 Find the pair of equations that do not represent the same graph:

a $3x + y = 12$ **b** $4x + 8y = 16$ **c** $3y - 6x = 9$ **d** $2y - 4x - 3 = 0$

$y = 12 - 3x$ $y = 2 - \frac{1}{2}x$ $y = 2x + 6$ $y = 2x + 1\frac{1}{2}$

e For the graphs in the odd pair, give their equivalent implicit or explicit form.

3 **a** Plot $2x + 3y = 12$ and $2x - y = 4$ on the same set of axes.

b Using your graph, solve:

$2x + 3y = 12$ and $2x - y = 4$ simultaneously.

4 Here are two rectangles:

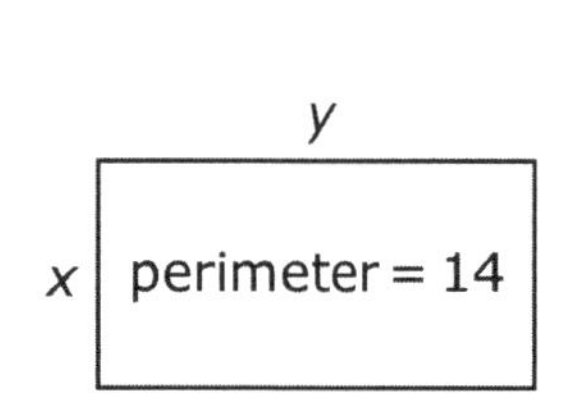

$2y + x$

perimeter = 26

$2x$

All lengths are given in cm.

a Write an equation for each rectangle to represent the perimeter.

b Plot these equations on a graph.

c Use your graph to find the dimensions of each rectangle.

Level 6

hese straight line graphs all pass through the point (10, 10).

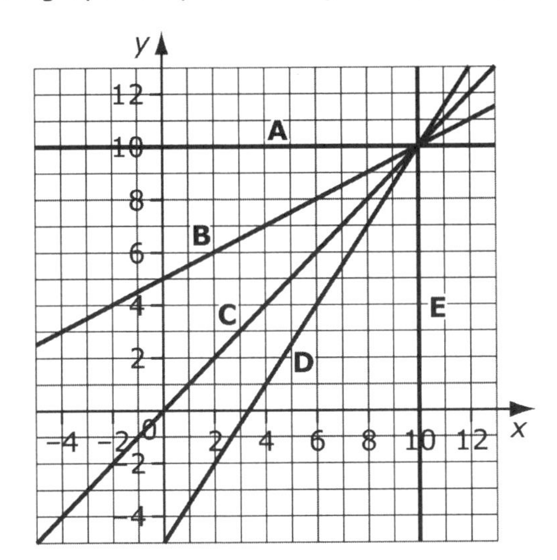

a Copy these sentences and fill in the gaps to show which line has which equation.

line .... has equation $x = 10$

line .... has equation $y = 10$

line .... has equation $y = x$

line .... has equation $y = \frac{3}{2}x - 5$

line .... has equation $y = \frac{1}{2}x + 5$ *2 marks*

b Does the line that has the equation $y = 2x - 5$ pass through the point (10, 10)?

Explain how you know. *1 mark*

c I want a line with equation $y = mx + 9$ to pass through the point (10, 10).

What is the value of m? *1 mark*

Level 7

a Two of the expressions below are **equivalent**.

Write them down.

$5(2y + 4)$ $5\ (2y + 20)$ $7(y + 9)$

$10(y + 9)$ $2(5y + 10)$ *1 mark*

b One of the expressions below is **not** a correct factorisation of $12y + 24$.

Which one is it?

$12\ (y + 2)$ $3\ (4y + 8)$ $2\ (6y + 12)$

$12\ (y + 24)$ $6\ (2y + 4)$ *1 mark*

c Factorise this expression.

$7y + 14$ *1 mark*

d Factorise this expression as fully as possible.

$6y^3 - 2y^2$ *2 marks*

P1.1HW **Understanding the problem**

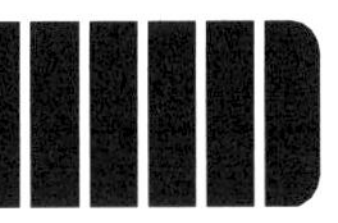

Calendar numbers investigation 1

Celia took a page of her calendar and drew a box around 4 numbers.

JUNE 2003						
SUN	MON	TUE	WED	THURS	FRI	SAT
1	2	3	4	5	6	7
8	9	10	11	12	13	14
15	16	17	18	19	20	21
22	23	24	25	26	27	28
29	30					

Celia calculated the products of the numbers in opposite corners of the box.

3	4
10	11

$4 \times 10 = 40$

$3 \times 11 = 33$

The difference between these products is 7 (= 40 – 33).

1 Investigate the difference between the products of opposite corners when you draw a box around 4 numbers on this calendar.

2 Investigate the difference between the products of opposite corners in boxes of 4 numbers for other arrays of numbers.

1st array

1	2
3	4
5	6
7	8

2nd array

1	2	3
4	5	6
7	8	9
10	11	12

3rd array

1	2	3	4
5	6	7	8
9	10	11	12
13	14	15	16

...

Remember to start with the simplest array and build up.

P1.2HW Organising your results

Calendar numbers investigation 2

Look at your results for the Calendar Number Investigation in Homework P1.1HW.

JUNE 2003						
SUN	MON	TUE	WED	THURS	FRI	SAT
1	2	3	4	5	6	7
8	9	10	11	12	13	14
15	16	17	18	19	20	21
22	23	24	25	26	27	28
29	30					

1 Draw a graph to show the difference in the products of opposite corners of a box around 4 numbers for different sized arrays of numbers. (Put array size on the horizontal axis and product on the vertical axis.)

2 Comment on the graph you have drawn.

3 Predict what the product of opposite corners would be for an array size that you have not yet drawn.

4 Test your prediction by drawing the array and a box around 4 numbers and calculating the product.

5 Find a general result for the product of opposite corners of a box with 4 numbers in an array with n numbers in a row.

Use algebra to explain why your rule always works.

P1.3HW Extending your investigation

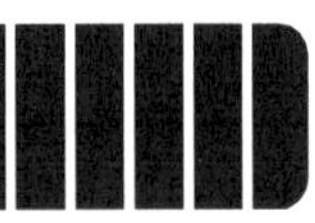

alendar number investigation 3

xtend your Calendar Numbers Investigation.

ou could ...

. draw different sizes of square boxes in an array of numbers.

. draw different sizes of rectangular boxes in an array of numbers.

. draw the same size rectangular box in different arrays of numbers.

emember:

- Do not try to extend your investigation in more than one way.
- Start with the simples case and increase in complexity by only changing one thing at a time.

ou should:

- Draw tables and diagrams to illustrate your results.
- Explain any patterns you can see from your tables and diagrams.
- Make predictions about future results and say why you think they are correct.
- Test your predictions.
- Try to find a general result.

Investigating ratios

The following sequence of numbers is the Fibonacci sequence:

1 1 2 3 5 8 13 21 34 ...

In the Fibonacci sequence the first two numbers are 1 and 1 and every subsequent number is equal to the sum of the previous two numbers; so the third number is 1 + 1 = 2, the fourth number is 1 + 2 = 3, and so on.

When you write adjacent numbers in the Fibonacci sequence as a ratio you get the following results:

1 : 1	1 : 2	2 : 3	3 : 5	5 : 8	...
First ratio	Second ratio	Third ratio	Fourth ratio	Fifth ratio	

Write each of these ratios in the unitary form 1 : n

Draw a graph to show your results.
Use axes like these.

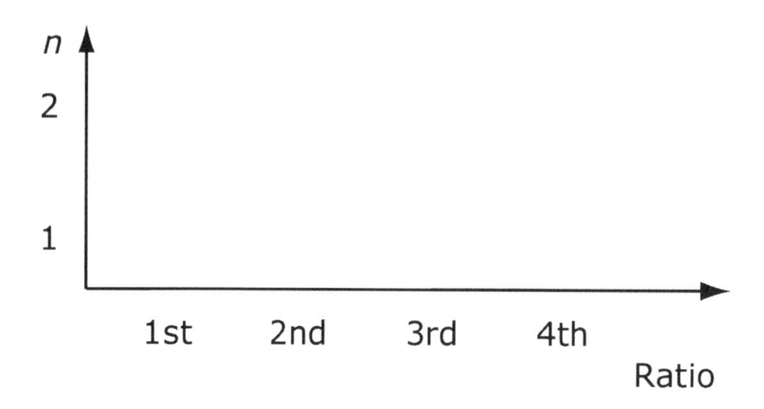

Use unitary ratios to investigate what happens to the ratio of adjacent numbers in the Fibonacci sequence.

P1.5HW Proportional reasoning

Molly makes honey cakes and honey biscuits to sell at country fairs.

Here are the recipes she uses.

Honey Cake (makes 1 cake)

140 g butter
120 g soft brown sugar
200 g self raising flour
150 g clear honey

Honey Biscuits
(makes a batch of 6 large biscuits)

70 g butter
30 g soft brown sugar
150 g self raising flour
60 g clear honey

These are the costs of each of the ingredients.

Butter	75p for 250 g
Soft brown sugar	£2.00 for 1 kg
Self raising flour	80p for 1 kg
Honey	£2.70 for 450 g

Work out the cost of making one honey cake and a batch of biscuits.

Molly wants to make a minimum profit of 25% when she sells the cakes and biscuits. How much should Molly sell them for at the country fairs? (Ignore the fuel cost for the baking.)
Round your answers to sensible amounts and give reasons to justify your rounding.

Molly actually sells each honey cake for £2.75 and each biscuit for 20p.

It takes Molly 20 minutes to make one cake and 15 minutes to make a batch of biscuits.

Which of the cakes and biscuits is it more profitable for Molly to make? Give reasons for your answers.

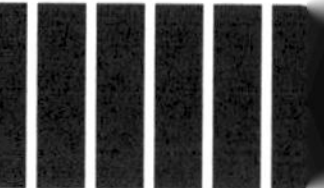

Morse code

In 1838 Samuel Morse devised a system of dots and dashes to represent letters.

This allowed short and long electric signals to be sent along wires to be translated into letters of the alphabet.

Find out:

- when Morse sent his first message
- what that message said.

Here are some places to look for information:

- http://www.connectedearth.com/
- http://www.radio-electronics.com/info/radio_history/morse/morsetelegstry.htm
- www.marconicalling.com
- http://www.porthcurno.org.uk/instrumentRoom/MorseKey.html
- Encyclopaedia or reference book.

Write a paragraph explaining how the Morse key is used to transmit messages.

Morse also devised a code for numbers.
Find and write down the code for each of the numbers 0 to 9.

Finish by writing out this sentence in Morse code:

In 1838 Samuel Morse devised Morse code

Solving problems

Level 6

Ann, **B**en, **C**arl, **D**onna and **E**ric are friends.

They have **four** tickets for a concert.
The **order** in which they sit at the concert **does not matter**.

Ben **must** go to the concert.
If **E**ric goes to the concert, **D**onna must go too.

List all the **different groups of four** who could go to the concert.

Remember, order does not matter. *2 marks*

Level 7

a $\boldsymbol{m}$ is an **odd** number.

Which of the numbers below must be even, and which must be odd?

Copy them down and write 'odd' or 'even' under each one.

$2m$	m^2	$3m - 1$	$(m - 1)(m + 1)$

2 marks

b $\boldsymbol{m}$ is an **odd** number.

Is the number $\frac{m + 1}{2}$ odd, or even, or is it not possible to tell?

Explain your answer. *1 mark*

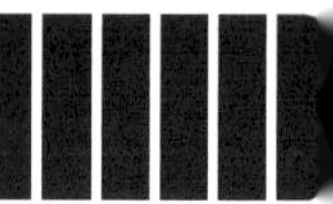

S4.1HW 3-D shapes

Find six household objects and describe their mathematical properties.

> **Hint:**
> Look for:
> - symmetry
> - number of faces, vertices and edges
> - parallel edges
> - perpendicular edges

For example,

a can of beans

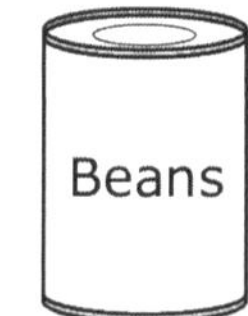

a pencil

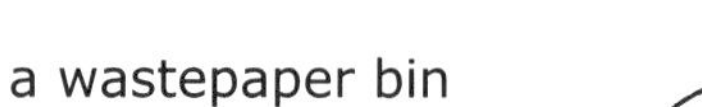

a wastepaper bin

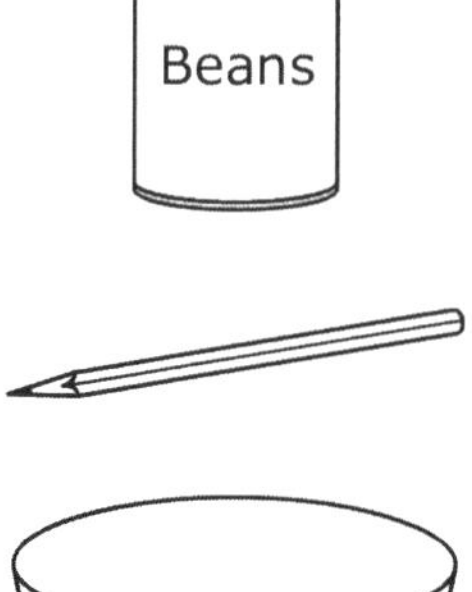

In each case, describe the object mathematically using terms such as cuboid, cylinder or cone as appropriate.

Write your results in a table like this:

Sketch	Object	Properties

S4.2HW Plans and elevations

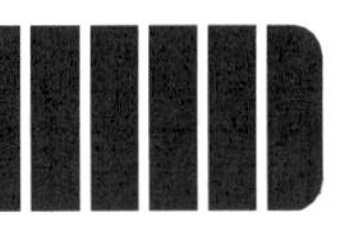

By making appropriate measurements, draw the nets of the solids shown in these plans and elevations.
Sketch each solid first.

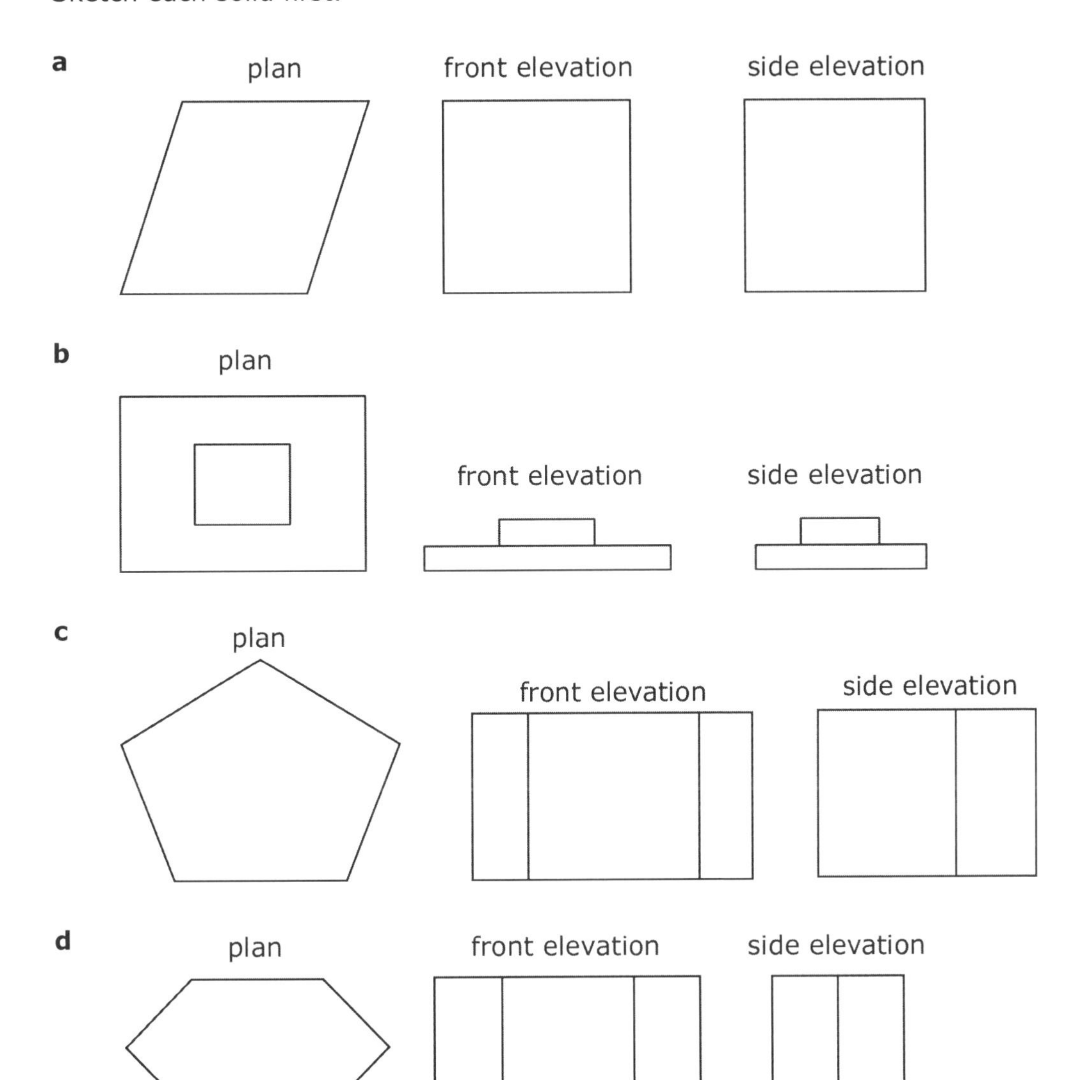

S4.3HW Volume and surface area

1 A plastic beaker has a height of 12 cm and a circular base of diameter 8 cm.

a Calculate the volume of the beaker.

b A label covers **all** the curved surface area. What is the area of the label?

C

12 cm

8 cm

Remember: $C = \pi d$ or $2\pi r$

2 This is the diagram of an Olympic sized swimming pool (not drawn to scale). Calculate the volume of the pool in m^3, and the surface area of the sides and base, which need painting.

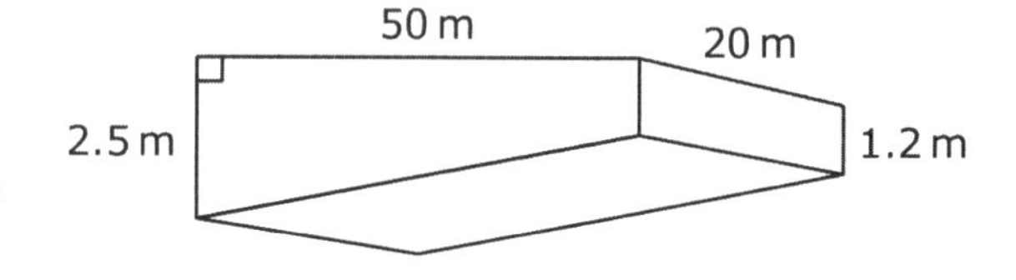

3 This net folds up to make a solid.

a What is the name of the solid?

b What is the volume of the solid?

c What is the surface area of the solid?

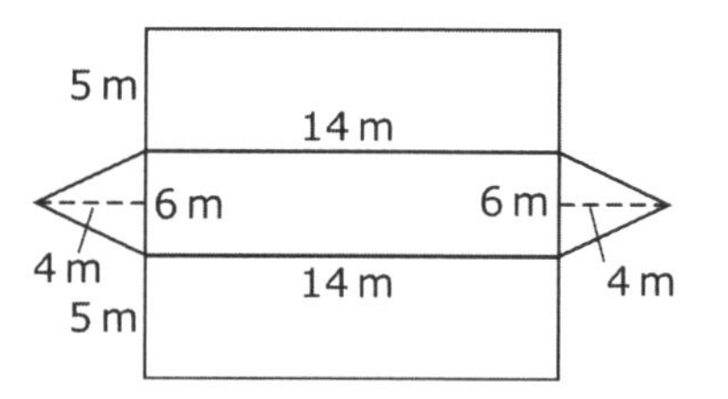

4 A circular can and a square based can both have the same height of 20 cm. The circular based can has a radius of 5 cm and the square based can has a side of length 10 cm. Both are full of soup.

Remember: Area of a circle $= \pi r^2$

a **i** Calculate volume of each type of can in cm^3.

ii Change these volumes to capacities in litres.

A school kitchen pan can hold 32 litres of soup.

b **i** How many square based cans will it take to fill the pan?

ii How many circular based cans will it take to fill the pan?

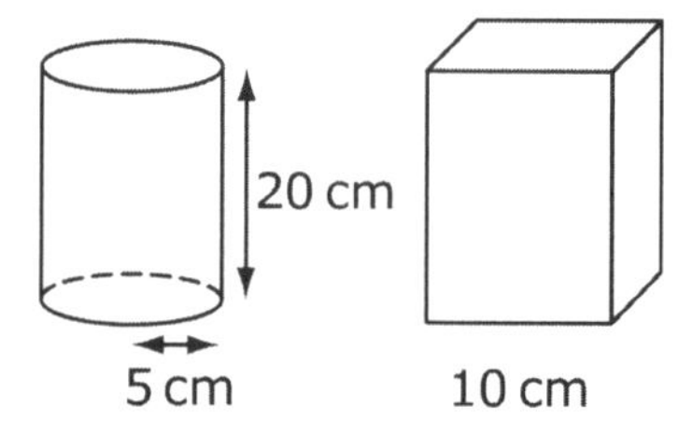

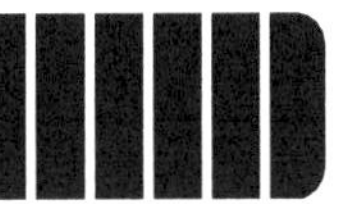

S4.4HW Scale drawings

Make a scale drawing of a room in your house.

- Mark on the doors and windows.
- Show the scale clearly.
- Include the furniture in the room, drawn to the correct scale.

For example, this scale drawing shows a kitchen.

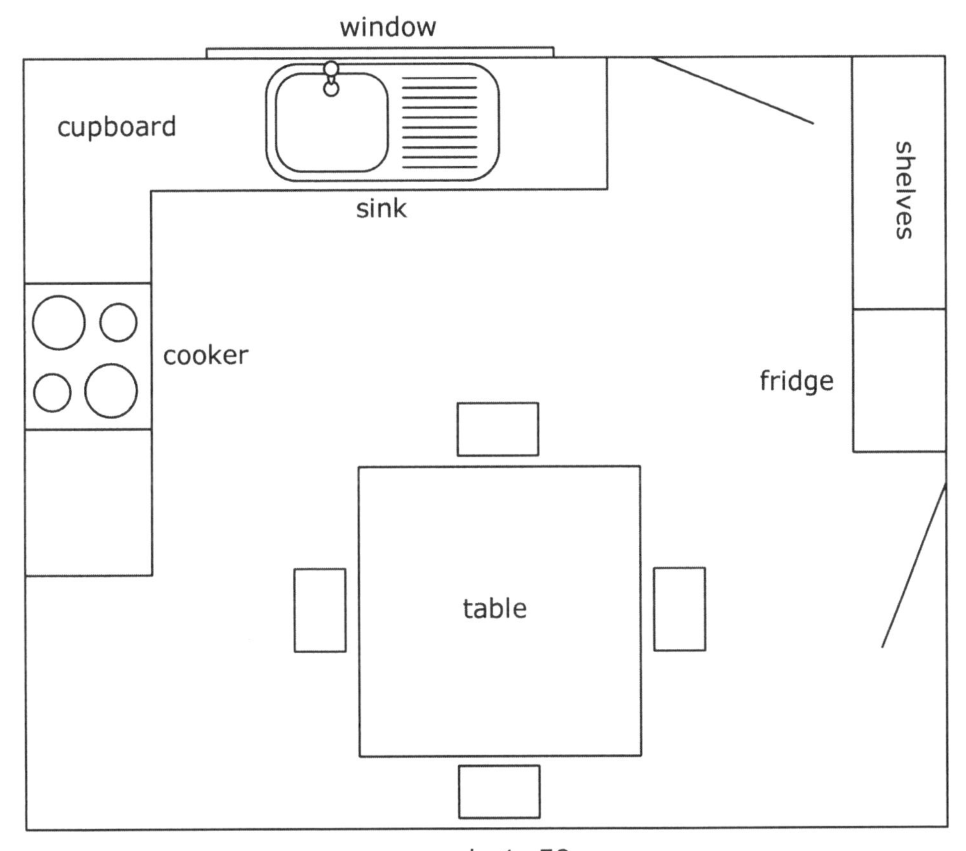

1 Measure the bearing of A from B in the following:

a **b** **c**

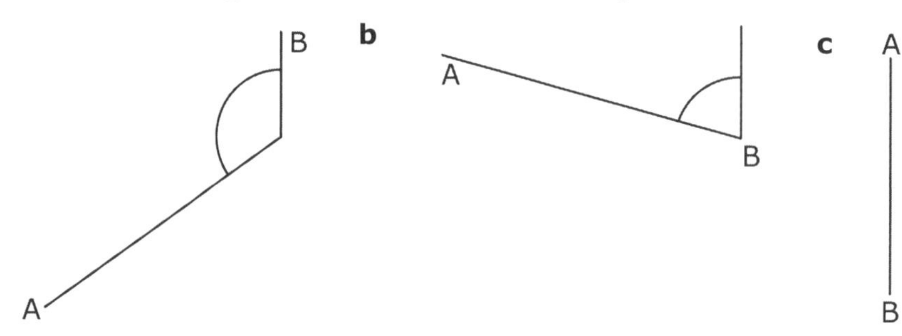

d **e** **f**

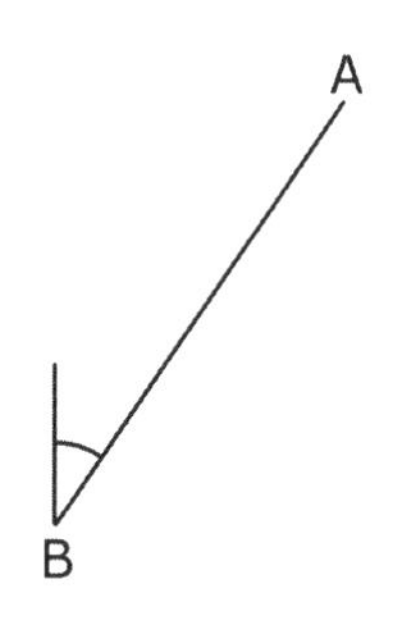

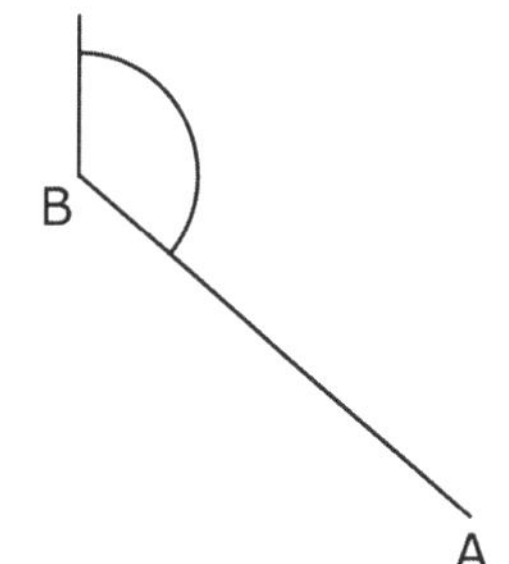

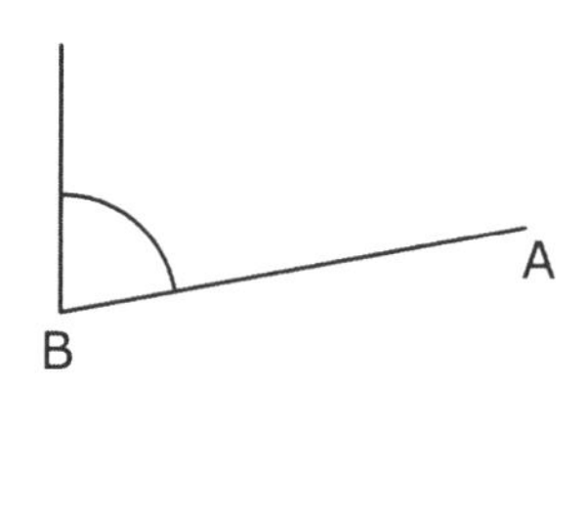

2 Measure the bearings of B from A in question 1.
What do you notice?

3 The bearing of Y from X is 059°.
What is the bearing of X from Y?

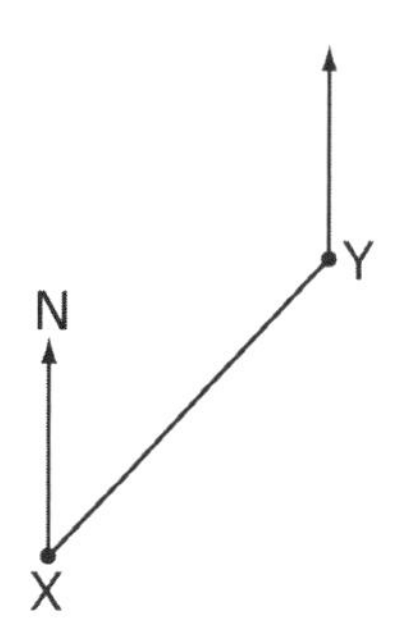

4 Tosca is on a bearing of 120° from the boat and 240° from the lighthouse. Make an accurate scale drawing to show its location. Find *d* and *e*.

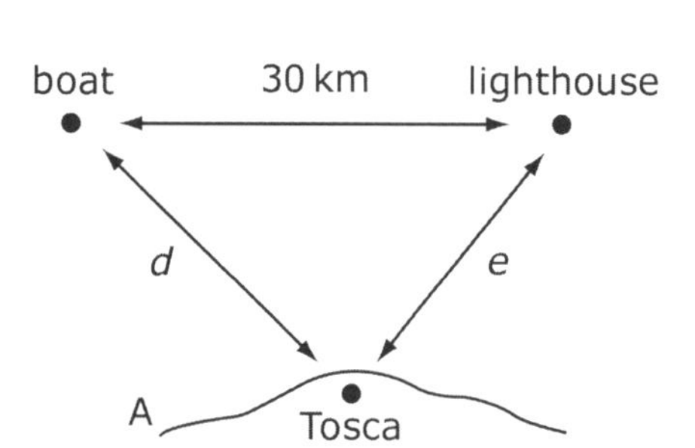

S4.6HW The midpoint of a line

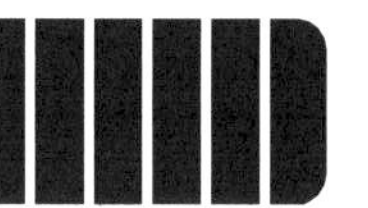

1. Find the midpoint M of the lines joining these pairs of points.

   a A (2, 3) B (4, 7)

   b C ($^{-}1$, $^{-}3$) D (2, 4)

   c E ($^{-}1$, 5) F ($^{-}6$, 10)

   d G (0, 4) H (0, $^{-}3$)

2. For a line AB, if A is (2, 3) and the midpoint M is (5, 7), find B.

3. For a line AB, if A is ($^{-}2$, $^{-}4$) and the midpoint M is ($^{-}1$, $^{-}2$), find B.

4. For a line AB, if A is (0, 2) and the midpoint M is (0, 6), find B.

5. **a** Draw the shape with coordinates A (2, 3), B (4, 5), C (9, 5) and D (7, 3) using suitable axes.

   b What is the mathematical name of the shape ABCD?

   c Find the coordinates of the midpoint of AC.

   d Find the coordinates of the midpoint of BD.

   e What do you notice from your answers to parts **c** and **d**?

S4.7HW Finding a simple loci

The diagram shows a plan view of a hut.

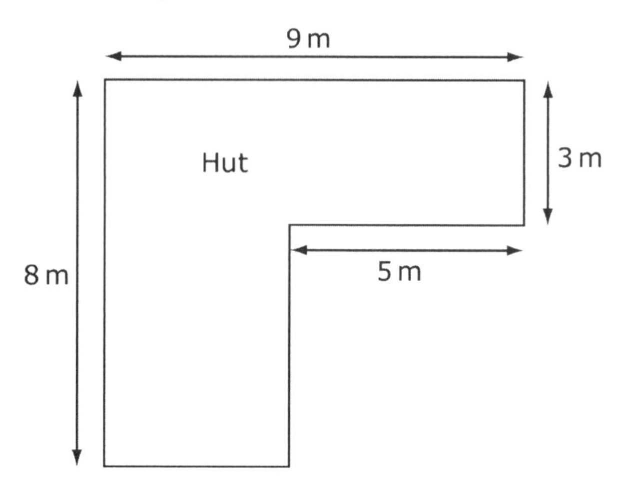

A robot is programmed so that it is always 1 m away from the walls of the hut.

Hint:
The robot can be inside or outside the hut.

1 Draw a scale diagram of the hut. On it show the locus of the robot both inside and outside.

2 Calculate the area contained between the two loci.

S4.8HW More loci

1 Copy points A and B. Construct the locus of a dog that walks on a path equidistant from A and B.

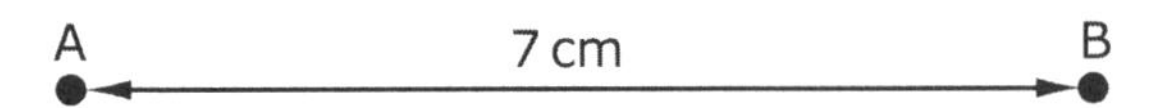

2 An ant walks so that it is always 4 cm from a piece of bread. Sketch the locus of the path of the ant.

3 **a** Construct a plan of the field PQRS using a scale of 1 : 500 (1 cm : 5 m)

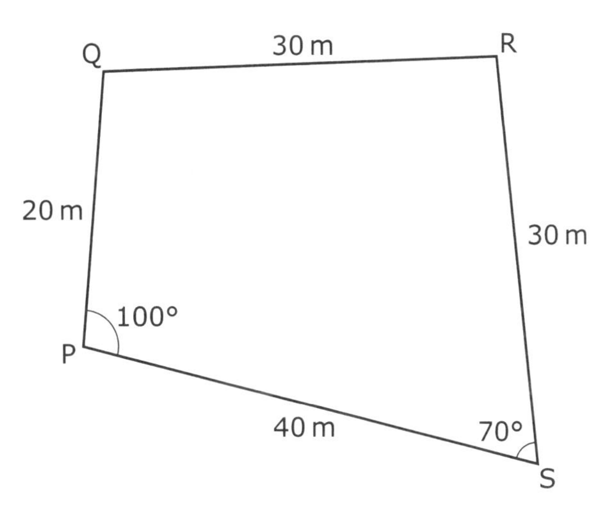

b A horse walks so that its path is equidistant from PQ and PS. Draw the locus of the horse's path accurately on your plan.

c A farmer plants apple trees so that they are all 15 m from S. Can the horse reach an apple tree?

S4.9HW Constructing triangles

1 The prince can see Rapunzel at an angle of elevation of 60°. How far is he from the base of her tower (d)?

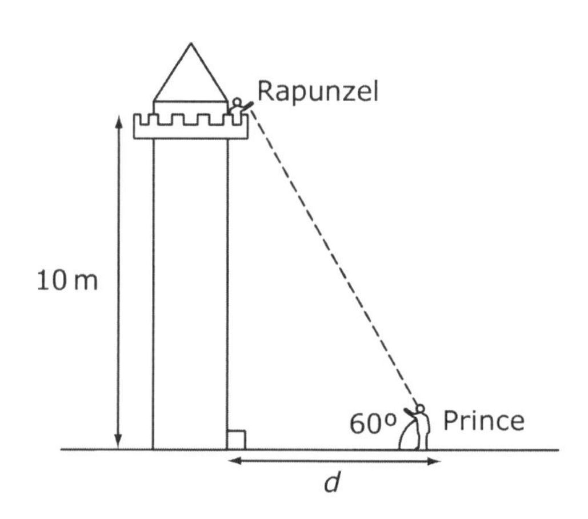

2 Construct a scale diagram to find the angle (θ) that the submarine needs to turn through to head towards the lighthouse instead of the tanker.

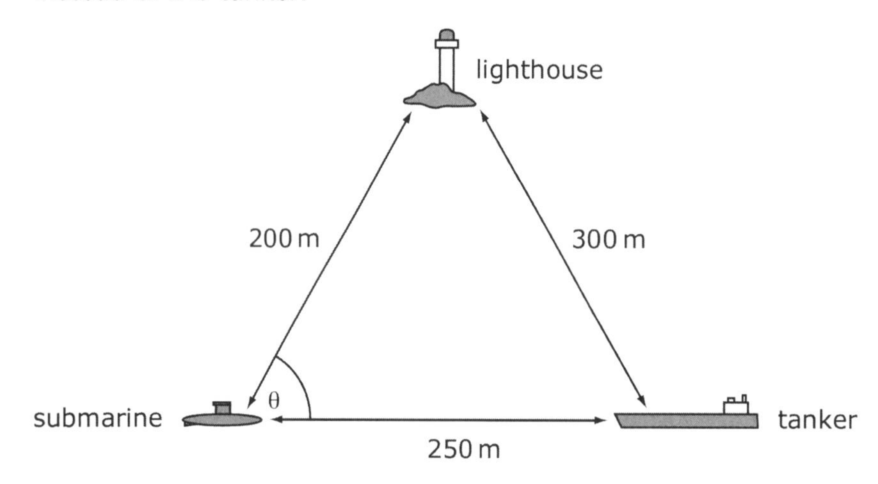

3 Is it possible to construct $\triangle ABC$ such that:

a BC = 6 cm, AC = 4 cm, AB = 3 cm

b BC = 7 cm, AC = 3 cm, AB = 2 cm

c $\angle A = 30°$, $\angle B = 45°$, AC = 6 cm

d $\angle A = 80°$, $\angle B = 90°$, $\angle C = 20°$

e BC = 7 cm, AC = 5.5 cm, $\angle B = 45°$

f BC = 7 cm, AC = 4 cm, $\angle B = 45°$?

Give reasons for your answers, and construct the triangle where possible.

Level 6

The diagram shows a model made with **nine** cubes.

Five of the cubes are grey. The other four cubes are white.

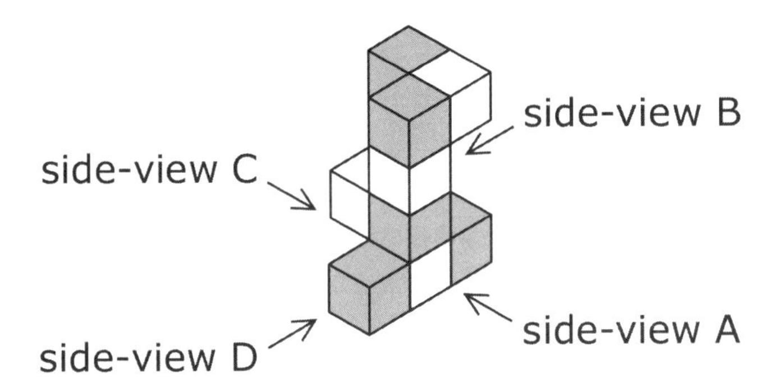

a The drawings below show the four side-views of the model.

Which side-view does each drawing show?

i

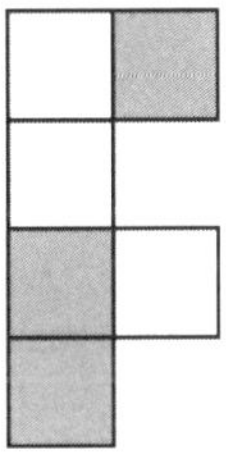

ii

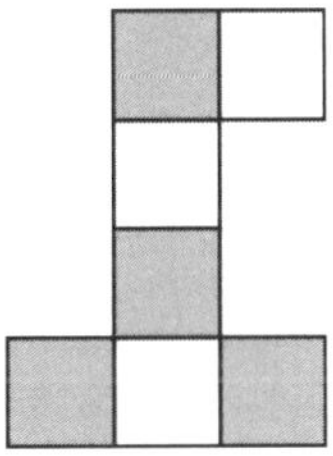

iii

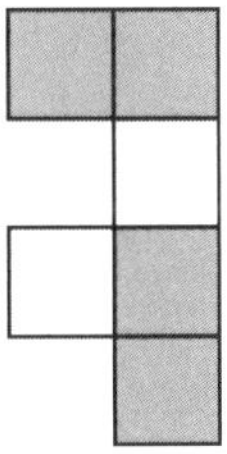

iv

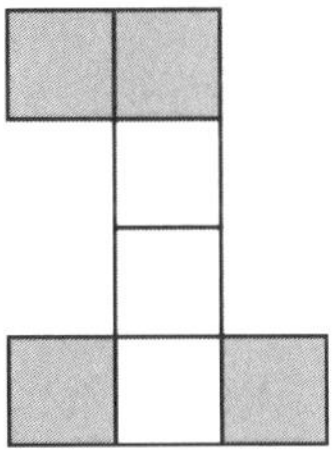

1 mark

continued

b Copy and complete the **top-view** of the model by shading the squares which are **grey**.

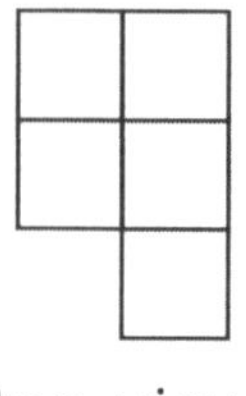

top-view

1 mark

c Imagine you turn the model **upside down**.

What will the new top-view of the model look like?

Copy and complete the **new top-view** of the model by shading the squares which are **grey**.

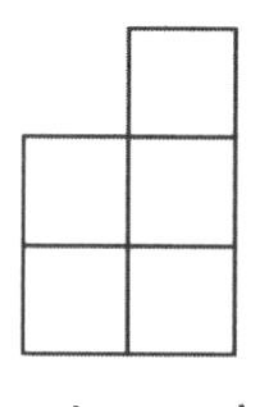

new top-view

1 mark

Level 7

The diagram shows the locus of all points that are the **same distance** from A as from B.

The locus is one straight line.

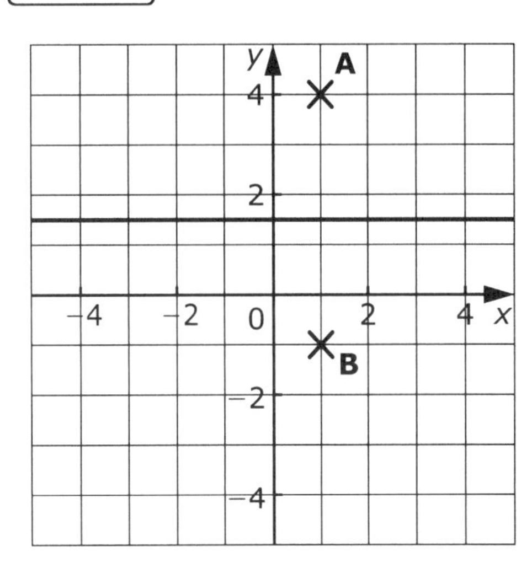

a The locus of all points that are the **same distance** from (2, 2) and (− 4, 2) is also one straight line.

Draw this straight line on a copy of the grid. *1 mark*

b The locus of all points that are the **same distance** from the x-axis as they are from the y-axis is **two** straight lines.

Draw both straight lines on a copy of the grid.

1 mark

c Look at points C and D below.

Copy the diagram. Use a straight edge and compasses to draw the locus of all points that are the same distance from C as from D.

Leave in your construction lines.

•C

•D

2 marks

D3.1HW Formulating a hypothesis

Music survey

Many people are now listening to popular music from twenty to thirty years ago.
For example:

- The music of Queen is the basis of the musical, 'We Will Rock You'.
- The musical 'Our House' features the music of Madness.
- Many of the songs featured in the film 'Billy Elliot' are from T-Rex.

You are going to investigate music originally produced twenty to thirty years ago and music produced more recently.

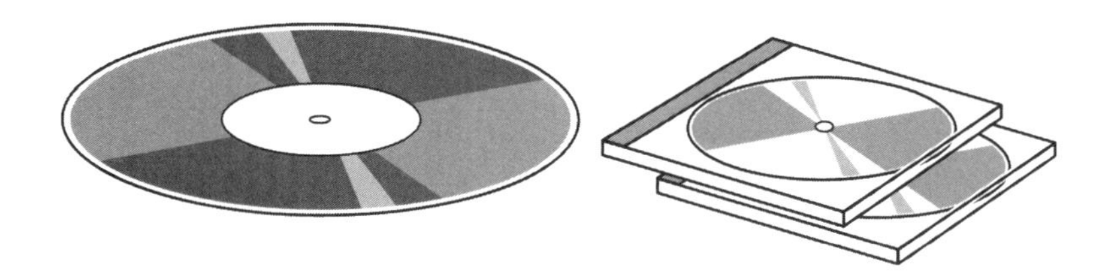

- Write out a hypothesis that you would like to investigate. Make sure it includes a comparison between music of different decades.
- Identify what data you will need to test your hypothesis. State if it is primary or secondary, and where you might find your data.

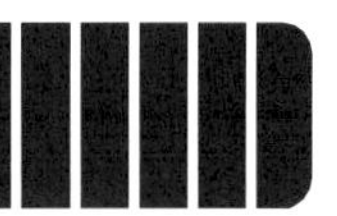

Music survey 2

Using the hypothesis that you have already written for the homework D3.1:

1. Describe how you would carry out a survey to collect this data.

   Remember to include detail on how much data you need to collect and how accurate you need to be.

   You need to think about these questions:

   - What or who will you survey?
   - What is your sample size?
   - What information to you need to collect?

2. Design a data collection sheet to collect the data.

   You need to think about these questions:

   - What headings will you need on your columns?
   - How many rows and columns will you need?
   - Is it easy to use?
   - Will it enable you to answer your hypothesis?
   - Will it enable you to sort your data easily?

Music survey 3

Here are the lengths of tracks, in minutes and seconds, on each of two CDs:

- music from the seventies, T-Rex
- music from the nineties, Boyzone.

T-Rex							
3:34	4:29	3:47	2:18	4:16	5:00	2:32	2:30
2:19	3:43	3:27	3:11	5:02	4:00	2:11	2:52
2:49	4:33	3:19	2:57	2:24	3:40	1:46	2:22

Boyzone						
3:42	4:34	3:46	3:15	3:29	3:45	4:15
4:04	2:46	3:03	3:45	3:45	3:39	3:37
3:29	4:18	4:14	4:10	3:29		

- Use these data to calculate statistics and draw diagrams that will answer the hypothesis:

  'Tracks from the nineties are shorter than tracks from the seventies.'
- Collect some data of your own and use both sets of data to answer your hypothesis from homeworks D3.1 and D3.2.

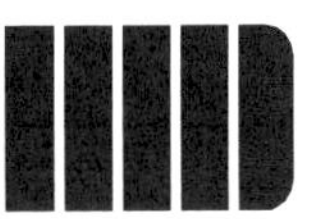

D3.4HW Interpreting and discussing data

Music survey 4

These graphs were drawn to represent the lengths of tracks of music for the data collected from the seventies and nineties in homework D3.3:

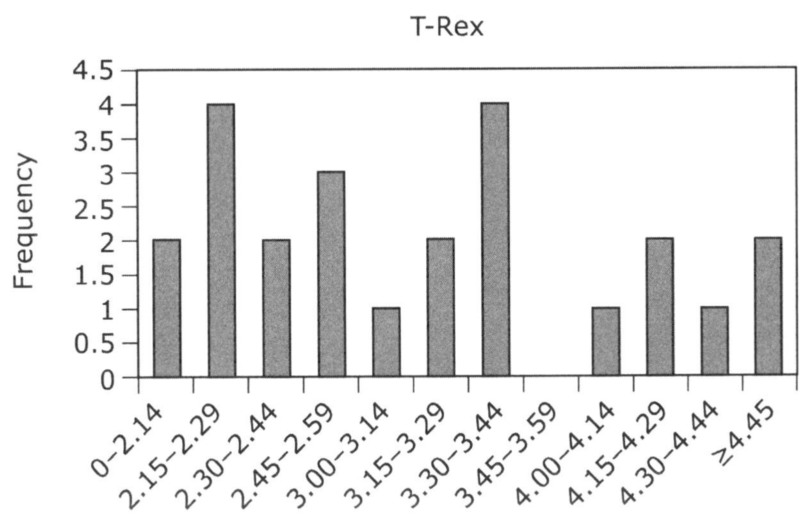

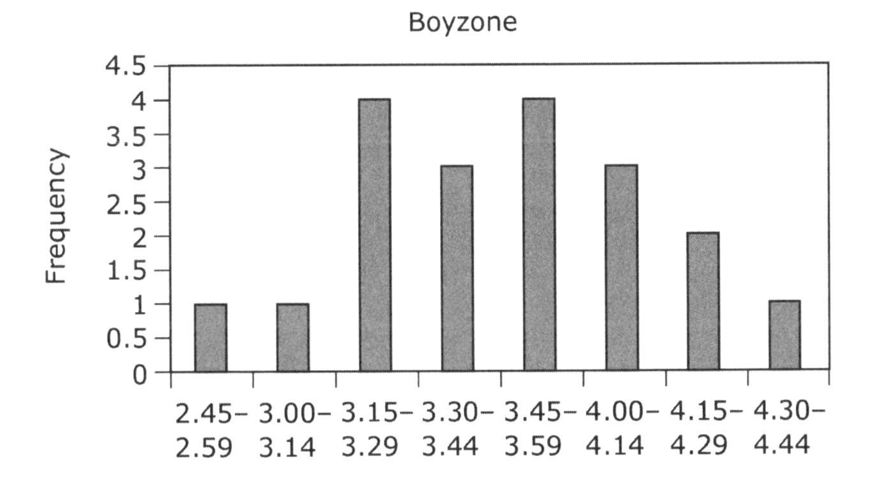

Comment on the trend shown by these two graphs.
Write down any similarities and differences in music length that you can see in the decades that these graphs represent.

Use your own calculated graphs and statistics to give a fuller answer.
You can also use your graphs from homework D3.3.

Level 6

A newspaper wrote an article about public libraries in England and Wales. It published this diagram.

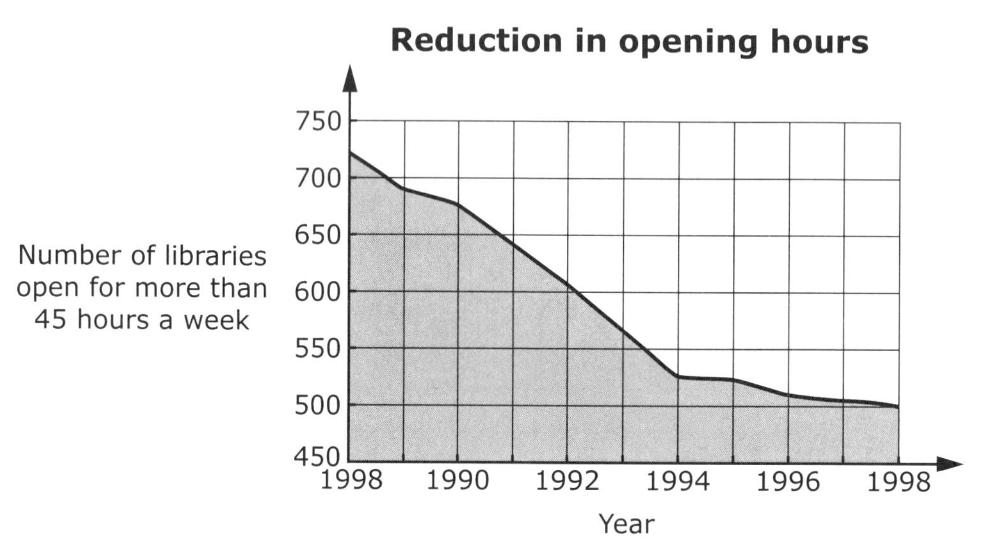

Data on libraries from LISU (Library and Information Statistics Unit)

Use the diagram to decide whether each statement below is true or false, or whether you cannot be certain.

a The number of libraries open for more than 45 hours per week **fell by more than half** from 1988 to 1998.

◆ True ◆ False ◆ Cannot be certain

Explain your answer. *1 mark*

b **In 2004** there will be **about 450 libraries** open in England and Wales for more than 45 hours a week.

Explain your answer. *1 mark*

Level 7

The percentage charts show information about the wing length of adult blackbirds, measured to the nearest millimetre.

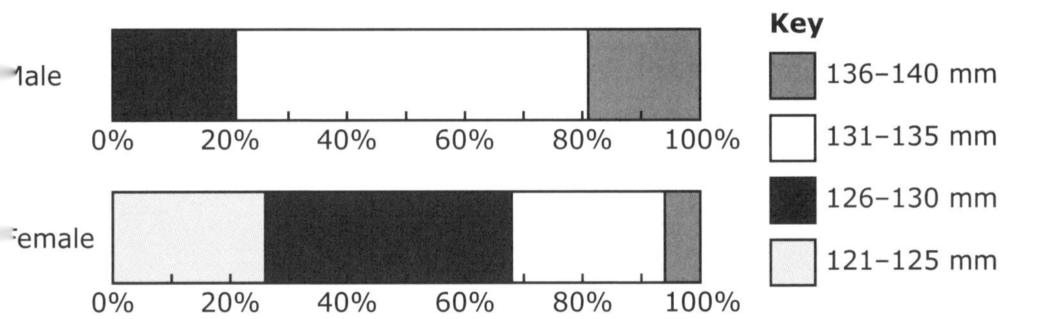

a Use the data to decide whether these statements are true or false, or whether there is not enough information to tell.

The smallest male's wing length is larger than the smallest female's wing length.

◆ True ◆ False ◆ Not enough information

Explain your answer. *1 mark*

b *The biggest male's wing length is larger than the biggest female's wing length.*

◆ True ◆ False ◆ Not enough information

Explain your answer. *1 mark*

Charity game

A roll-a-coin charity game is set up at a fete.
10 pence coins are rolled onto a board.
The board is a grid of 25 squares. Each square has side length 40mm.

You win if the 10 pence coin touches the winning square in the centre of the board.

A 10 pence piece has radius about 25mm.

1 If the grid lines on the board have no thickness, what is the probability that a coin rolled onto the board lands touching the winning square?

2 If the grid lines on the board have a thickness of 2mm, what is the probability that a coin rolled onto the board lands touching the winning square?

3 Make a grid of 25 squares where each square has side length 40mm. (You will need paper, card or a wood that is at least 20cm × 20cm).

- Roll a 10 pence piece onto the grid twenty times.
- Compare your experimental results with the theoretical results you have already calculated.
- Which of the two theoretical results matches your experiment best?

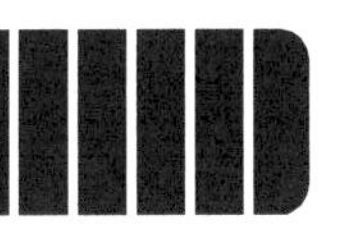

Dolphin Derby

Dolphin Derby is a game played at the end of a pier.

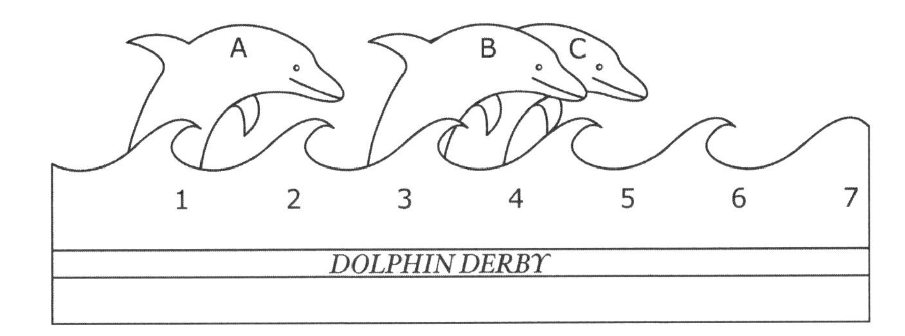

Each player is assigned a dolphin, A, B or C.

Each player has 3 balls that they roll onto a board.

The board has holes into which the balls drop.

Each hole is worth a specified number of points.

Each time a ball falls through a hole, the dolphin moves forward that number of paces.

The first dolphin to reach the end at 15 points is the winner.

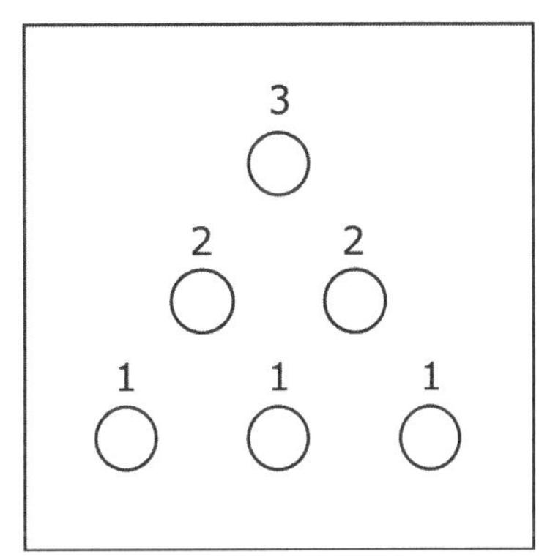

◆ With a friend or in a small group simulate playing the game, 'Dolphin Derby'.

D4.3HW Statistical experiments

You need to carry out a large number of trials when you conduct a statistical experiment.

Collating all the data you collect and doing all the calculations would take a long time without using computers and calculators.

Charles Babbage invented the world's first digital computer, the mechanical 'Analytical Engine'.

Modern calculators are based on his inventions.

Ada Lovelace worked with Charles Babbage and wrote the first computer program.

Find out:

- when The Analytical Engine was built
- when the first computer program was written.
- Describe how the Analytical Engine works.
- List some of the differences between this first computer and modern-day calculators.

Here are some places to look for information:

- www.devon.gov.uk/babbage/banal.html
- www.fourmilab.ch/babbage/contents.html

Level 6

have two fair dice.

:ach of the dice is numbered 1 to 6.

a The probability that I will throw **double 6** (both dice showing number 6) is:

$$\frac{1}{36}$$

What is the probability that I will **not** throw double 6? *1 mark*

b I throw both dice and get double 6.
Then I throw the dice again.

Write down the statement that describes the probability that I will throw **double 6** this time:

- less than $\frac{1}{36}$
- $\frac{1}{36}$
- more than $\frac{1}{36}$

Explain your answer. *1 mark*

I start again and throw both dice.

c What is the probability that I will throw **double 3** (both dice showing 3)?

1 mark

d What is the probability that I will throw a double?
(It could be double 1 or double 2 or any other double.)

1 mark

Level 7

a Alan has a guessing game on his computer.

He estimates that the probability of **winning** each game is **0.35**

Alan decides to play **20** of these games.

How many of these games should he expect to **win**? *1 mark*

b Sue played the same computer game.

She won **12** of the games she played, and so she estimated the probability of winning each game to be **0.4**

How many games did Sue play?

Show your working. *2 marks*

c The manufacturers of another guessing game claim that the probability of winning each game is **0.65**

Karen plays this game **200** times and **wins 124** times.

She says: 'The manufacturers must be wrong'.

Do you agree with her? Write Yes or No.

Explain your answer. *1 mark*